AF412903

BioValley Monographs

Vol. 3

Series Editors

Philippe Poindron Strasbourg
Pascale Piguet Basel

This book is sponsored by

Conseil Régional d'Alsace 

Endress Foundation

BioValley Alsace

Genetically Modified Organisms and Genetic Engineering in Research and Therapy

Volume Editors

Pascale Piguet Basel
Philippe Poindron Strasbourg

6 figures, 4 in color and 6 tables, 2012

Basel · Freiburg · Paris · London · New York · New Delhi · Bangkok · Beijing · Tokyo · Kuala Lumpur · Singapore · Sydney

BioValley Monographs

Pascale Piguet
PharmaZentrum
Klingelbergstrasse 50
4050 Basel
Switzerland

Philippe Poindron
11 rue Galliéni
92100 Boulogne
France

Library of Congress Cataloging-in-Publication Data

Genetically modified organisms and genetic engineering in research and
therapy / volume editors, Pascale Piguet, Philippe Poindron.
     p. ; cm. -- (BioValley monographs, ISSN 1660-8984 ; v. 3)
  Includes bibliographical references and indexes.
  Summary in English and French.
   ISBN 978-3-8055-9065-5 (hard cover : alk. paper) -- ISBN 978-3-8055-9066-2
(e-ISBN)
   I. Piguet, Pascale. II. Poindron, Philippe. III. Series: BioValley
monographs ; v. 3. 1660-8984
   [DNLM: 1. Organisms, Genetically Modified. 2. Gene Therapy. 3. Gene
   Transfer Techniques. 4. Genetic Engineering. QU 450]
   615.8′95--dc23

                                              2012022409

Contents

V

Foreword

As with many revolutionary ideas, BioValley biocluster of the Upper Rhine Area, after an initial phase of enthusiastic support from the scientific, academic, economic and political world, has entered a phase of what, in French, is called 'un rythme de croisière'. Few impatient people still believe that it has not produced all the fruits expected from it and look upon it disappointedly. International crises and a fear of losing national specificity have certainly played a role in the stagnation of this initial, ambitious project. This is why it has taken so long to publish this, the third, issue of *BioValley Monographs.*

It would never have been published without the patient, untiring work of Dr. Pascale Piguet, the invaluable collaboration of Karger Publishers, and the financial support from the Région Alsace, Endress Foundation and the Alsace BioValley Association. At this point, it is impossible not to evoke the memory of Dr. Georg Endress, father of the BioValley initiative. He held most profound views on the direction the world was taking; he was generous, open-minded, and had always, both intellectually and financially, supported his 'child'.

The world has become a huge village. I think that the time has now come to combine the efforts of all European bioclusters to make the innovative potential of Europe more visible compared with that of the USA or China. This is why I proposed to transform the present series into *European Bioclusters Monographs.* I realize that to convince all the people in charge of these clusters will be an enormous task, but I would appreciate if all parties concerned tried their hardest to explore this path despite the intensive efforts needed for it to be successful.

Let me now gratefully acknowledge not only the contributors to the present issue, but also the institutions that gave financial support to its launching. First, the Région Alsace, the President of which, Adrien Zeller, died about 2 years ago. He approved and supported the present project. I dedicate this issue to his memory. Adrien Zeller was also a personal friend and I think that he would have been proud to tell me how happy he was to see the results of his ideas. Many grateful thanks go to the sons of Dr. Endress who provided financial support to the publication of this monograph via the Endress Foundation as well as the Alsace BioValley Association who also kindly supported us. Some years ago, I was president of this association which has now changed its status in the frame of the French 'Pôles de compétitivité'. I am happy to see that the present leaders did not forget that *BioValley Monographs* was launched when I was president and that I have never stopped believing in the value of their publication.

Philippe Poindron

Summary/Résumé

Genetically modified organisms (GMO) raise societal, political and ethical concerns. They inspire strong resistance or, conversely, enthusiastic assent. What problems are faced in the field of genetic modification, and what scientific questions do they encourage to be answered? The aim of this volume is to give an overview of genetic engineering from the technical aspects of transgenesis to its application in research. Be it through the use of individual transfected cells or GMO, these applications cover a broad array, ranging from disease-oriented research (but not only) to important therapeutic perspectives. Gene therapy is described in full-length articles devoted to the use of this technique in cancer and in muscular disease.

In the first section, we define the following concepts: organisms, modifications of germen and soma, vectors and vector expression, as well as transgene. Then, we briefly describe the methods being used currently to create GMO. A short description of the main transgenic organisms is given in the first chapter. We do not intend to be exhaustive, but demonstrative in the hope of providing our readers with objective elements to help them form their own opinions.

In the second chapter, Prof Werner Arber describes the history that led to the discovery of the molecular mechanisms of restriction and modification that control the success of the infection of bacteria by viruses called bacteriophages. Study of these – first unraveled – epigenetic phenomena led to the isolation of bacterial endonucleases. Their wide use since then has had a tremendous impact on molecular genetic analysis and biotechnological innovations, including genetic engineering. Thanks to the restriction enzymes, much scientific knowledge has accumulated in many disciplines.

The use of GMO or genetically modified cells has played a significant role in elucidating some of the physiological and/or pathological functions of endogenous molecules. In the article by Spittau and colleagues, it is shown how the use of genetically modified mice has uncovered some key physiological functions of extracellular signaling molecule members of the TGF-β superfamily. Many of these execute important functions during early embryogenesis, organogenesis, after birth and in the adult, as well as in tissue repair and homeostasis.

The study of pathological mechanisms has also benefited from genetic engineering of cells or whole organisms. In their article, Dr. Patte-Mensah and colleagues explain how transfection of a human neuroblastoma cell line with molecular hallmarks of Alzheimer's disease was used as a strategy to uncover the role of neurosteroidogenesis in neuroprotection. *The technological approach used in this article appears as promising to obtain valuable insights into clarification of the pathophysiological mechanisms involved in neurodegenerative diseases.* Besides deciphering the pathological mechanisms involved in the development of a disease, GMO may participate in developing therapeutic strategies. In their article about the fragile X syndrome (the most common form of inherited intellectual disability and a leading cause of autism), Dr. Michalon and colleagues describe a mouse model, called Fmr1 knockout, which presents behavioral, anatomical and biochemical alterations similar to the ones observed in the human condition. As explained by these authors, *the availability of a rodent disease model with strong construct and face validity has been essential for our current understanding of the pathophysiology and remains an indispensable tool for testing therapeutic hypotheses.* It will be of great interest to assess whether the pharmacological findings obtained using these GMO will be applicable to human clinics. The hope to use genetic engineering as a way to treat a disease was already born a few decades ago. In the 2000s the famous success obtained by Prof. Alain Fischer in the treatment of severe combined immunodeficiency in children re-activated the hope of using gene therapy in hereditary as well as nonhereditary diseases. In their article on neuromuscular diseases, Dr. Cure and Dr. Braun detail the numerous challenges facing gene therapy. They describe which strategies are used in this field in order to achieve the progressive development of innovative therapeutic perspectives in animals and humans. Even though gene therapy is not indicated in clinical practice yet, it has been the object of intense research and trials. As thoroughly illustrated by Dr. Hajri, *significant advances have also been made in the field of gene therapy against cancer for the identification of new therapeutic genes, the design of innovative vectors and different ways of targeting during the last decades.* The author details the scientific evolution which has allowed us to actively overcome the past obstacles and barriers that have slowed down progress in this field.

Some further complexity when considering genetic information in eukaryotic cells results from the fact that such information is distributed between the nucleus and cytoplasmic organelles, in particular mitochondria. As explained by Dr. Dietrich and colleagues, *mitochondria ensure fundamental functions in energy production, redox processes, metabolic pathways and cell death. Their genetic system provides a number of polypeptides which are essential for cell survival, as they are components of the respiratory chain or contribute to its biogenesis. All data imply that maintenance, integrity and efficient expression of the mitochondrial genome are fundamental for eukaryotic organisms. In humans, mutations in the mitochondrial DNA are the most common cause of hereditary neuromuscular diseases. These degenerative disorders are still incurable and would need gene therapy. In plants, mitochondrial genetics provides traits of great agronomical relevance.* Thus, targeting mitochondrial genetic information might be of fundamental importance in the therapy of human disorders. *Conventional methodologies failed to enable genetic transformation of mitochondria in mammalian and plant cells, but a variety of alternative complementation strategies and organelle transfection approaches are currently developed.* What are these approaches and how may genetically modified plant cells contribute in the fight against human diseases? The authors address this problem in a final article which uncovers the possible biomedical applications of genetically modified plants, thus raising the broader question of the general use of GMO.

What about risks and safety? Prof. Arber explains how *the risks that could be inherent to the prospective field of genetic engineering were already discussed as early as 1972.* At that time, the Nobel laureate was *aware that a more profound scientific knowledge on molecular mechanisms of spontaneously occurring genetic variation, the driving force of biological evolution, was needed in order to compare natural genetic variation with engineered genetic variation.* Following further comparisons between natural evolution and designed steps of evolution by genetic engineering, the author concluded that *on the basis of ethical considerations, we, human beings, can ameliorate our bases and conditions of living as long as we fully respect the laws of nature and the established ethical norms. This also holds for microorganisms engineered to produce particular organic compounds or in view of their use for bioremediation. For ethical reasons, microorganisms should not be genetically engineered to render them stronger pathogens. Firm ethical norms should be followed for any deliberate genetic alteration in animals and in particular in human beings.* Such important ethical rules should remain in everybody's mind and remind us that any scientific discovery may become the best or the worst tool in human hands. Keeping this in mind, the *thereby acquired scientific knowledge on genomics, proteomics and molecular evolution also has its philosophical values and deserves to belong to the cultural heritage of mankind.*

Résumé

Les organismes génétiquement modifiés (OGM) soulèvent des questions de société, politiques et éthiques. Ils rencontrent une forte opposition ou, inversement, un agrément enthousiaste. A quels problèmes doit faire face le domaine de la modification génétique et quelles questions scientifiques contribue-t-il à élucider? Le but de ce volume est de donner une vue d'ensemble sur l'ingénierie génétique, allant des aspects techniques de la transgénèse à ses applications dans le domaine de la recherche. Que ce soit via l'utilisation de cellules transfectées individuelles ou d'OGM, ces applications couvrent un large éventail allant de la recherche, particulièrement axée sur le traitement des maladies (mais pas seulement), à d'importantes perspectives thérapeutiques. La thérapie génique est abordée en détail dans des chapitres dédiés à l'utilisation de cette technique pour le cancer et les maladies musculaires.

Dans une première section, nous définirons les concepts suivants: organismes, modification des cellules reproductrices et non reproductrices, vecteurs et vecteur d'expression, transgène. Nous décrirons ensuite brièvement les méthodes utilisées à l'heure actuelle pour créer des OGM. Une brève description des principaux organismes transgéniques sera donnée dans le premier chapitre. Il ne s'agit pas ici d'en donner une liste exhaustive, mais plutôt de faire un bref tour d'horizon et de fournir aux lecteurs des éléments objectifs afin de construire leur propre opinion.

Dans le second chapitre, le Professeur Werner Arber décrit l'histoire qui a mené à la découverte des mécanismes moléculaires de restriction et modification qui contrôlent le succès de l'infection des bactéries par des virus dits bactériophages. L'étude de ces phénomènes épigénétiques – alors mis en évidence pour la première fois – a conduit à l'isolation d'endonucléases bactériennes. Leur large utilisation depuis lors a eu un impact énorme pour les analyses de génétique moléculaire et pour les innovations biotechnologiques, y compris l'ingénierie génétique. Grâce aux enzymes de restriction, la connaissance scientifique a connu d'importants progrès dans de nombreuses disciplines.

L'utilisation de cellules ou d'organismes génétiquement modifiés a joué un rôle significatif dans l'élucidation de certaines fonctions *physiologiques* et/ou *pathologiques* de molécules endogènes. Dans le chapitre de Dr. Spittau et al., il est démontré comment l'utilisation de souris génétiquement modifiées a permis de révéler des fonctions-clés *physiologiques* de molécules de signalisation extracellulaire, membres de la superfamille TGF-β. Nombre d'entre elles remplissent d'importantes fonctions au cours de l'embryogenèse précoce, de l'organogenèse, après la naissance et chez l'adulte, de même qu'au cours de la réparation tissulaire et des processus d'homéostasie. L'étude des mécanismes *pathologiques* a également bénéficié de l'ingénierie génétique de cellules ou d'organismes entiers. Dans leur chapitre, le Dr. Patte-Mensah et al. expliquent comment la transfection d'une lignée cellulaire de neuroblastome humain à l'aide de molécules caractéristiques de la maladie d'Alzheimer a été utilisée comme stratégie pour dévoiler le rôle neuroprotecteur de la neurostéroïdogenèse. «*L'approche technologique utilisee dans ce chapitre apparait prometteuse afin d'obtenir des informations precieuses sur la clarification des mecanismes pathophysiologiques impliques dans les maladies neurodegeneratives.*» Les OGM sont capables de décoder les mécanismes pathologiques impliqués dans le développement d'une maladie, mais ils peuvent aussi contribuer au développement de stratégies thérapeutiques. Dans leur chapitre sur le syndrome dit du «X fragile» (la forme la plus courante de handicap intellectuel héréditaire et l'une des causes principales de l'autisme), le Dr. Michalon et al. décrivent un modèle de souris, dit «knock-out Fmr1», qui présente des altérations comportementales, anatomiques et biochimiques similaires à celles observées dans le syndrome humain. Ainsi que ces auteurs l'expliquent, *«la disponibilite chez le rongeur d'un modele de maladie avec une forte validite de concept a ete essentielle a notre comprehension actuelle de la pathophysiologie et reste un outil indispensable afin de tester des hypotheses therapeutiques»*. Il sera d'un grand intérêt d'évaluer à présent si les découvertes pharmacologiques obtenues à l'aide de ces OGM seront applicables à la pratique clinique chez l'humain. L'espoir d'utiliser l'ingénierie génétique comme outil de traitement des maladies naissait il y a déjà quelques décennies. C'est dans les années 2000, à la suite du succès retentissant obtenu par le Pr A. Fischer dans le traitement d'enfants atteints d'un déficit immunitaire combiné sévère, que l'espoir d'utiliser la thérapie génique dans les maladies héréditaires, mais également non héréditaires, avait été réactivé après de nombreux échecs. Dans leur chapitre sur les maladies neuromusculaires, les Drs. Cure et Braun détaillent les nombreux défis auxquels la thérapie génique fait face dans ce domaine. Ils décrivent quelles stratégies sont poursuivies afin de développer progressivement des perspectives thérapeutiques innovantes pour ces pathologies chez l'animal et l'homme. Bien que la thérapie génique ne soit pas encore indiquée dans la pratique clinique, elle a fait l'objet d'une recherche intense et de nombreux essais cliniques. Comme l'illustre en détail le Dr. Hajri, *«des avancees significatives ont egalement ete faites dans le domaine de la therapie genique contre le cancer, pour l'identification de nouveaux genes therapeutiques, la conception de vecteurs innovateurs et de differentes approches de ciblages au cours des dernieres decennies»*. L'auteur décrit l'évolution scientifique qui a permis peu à peu de vaincre les obstacles et barrières qui ralentissaient le progrès dans ce domaine.

Chez les eucaryotes, il existe un bagage génétique à la fois dans le noyau et les organelles cytoplasmiques, en particulier les mitochondries, ce qui ajoute à la complexité de ces investigations. Comme l'expliquent le Dr. Dietrich et al., *«les mitochondries assurent des fonctions fondamentales pour la production d'energie, les processus d'oxydo-reduction, les voies metaboliques et la mort cellulaire. Leur systeme genetique fournit un nombre de polypeptides essentiels a la survie cellulaire, du fait qu'elles sont des composantes de la chaine respiratoire ou contribuent a sa biogenese. Toutes les donnees impliquent que la maintenance, l'integrite et l'expression efficace du genome mitochondrial sont fondamentaux pour les organismes eucaryotes. Chez l'homme, certaines mutations dans l'ADN mitochondrial sont les causes les plus communes des maladies neuromusculaires hereditaires. Ces pathologies degeneratives sont encore incurables et necessiteraient une therapie genique. Chez les plantes, la genetique mitochondriale fournit des traits d'une grande pertinence agronomique»*. Ainsi, le ciblage de l'information génétique mitochondriale peut jouer un rôle fondamental dans la thérapie de pathologies humaines. *«Les methodologies conventionnelles ne permettent pas encore la transformation genetique des mitochondries dans les cellules des mammiferes et plantes, mais diverses strategies alternatives de complementation et approches de transfection des organelles sont developpees a l'heure actuelle.»* Quelles sont ces approches et comment des cellules de plantes génétiquement modifiées peuvent-elles contribuer à combattre des pathologies humaines? Les auteurs envisagent cette question dans un chapitre final qui dévoile les applications biomédicales éventuelles des plantes génétiquement modifiées, soulevant ainsi la question plus large de l'utilisation générale des OGM.

Qu'en est-il des risques et de la sécurité? Le Prof. Arber rappelle comment le problème des risques qui pourraient être inhérents au domaine prospectif de l'ingénierie génétique faisait déjà l'objet de discussions dès 1972. A cette époque, le prix Nobel était *«conscient du fait qu'une connaissance scientifique plus profonde des mecanismes moleculaires mis en oeuvre dans la variation genetique spontanee, la force motrice de l'evolution, etait necessaire afin de comparer la variation genetique naturelle et la variation genetique fabriquee par ingenierie».* Après d'autres comparaisons entre l'évolution naturelle et des paliers d'évolution conçus par ingénierie génétique, l'auteur conclut que *sur la base de «considerations ethiques, nous, en tant qu'etres humains, pouvons ameliorer nos bases et conditions de vie, tant que nous respectons pleinement les lois de la nature et les normes ethiques etablies. Ceci est egalement valable pour les micro-organismes concus afin de produire des composes organiques particuliers ou en vue de leur utilisation therapeutique. Pour des raisons ethiques, les micro-organismes ne devraient pas etre genetiquement modifies afin d'augmenter leur pouvoir pathogene. Des normes ethiques strictes doivent etre suivies pour toute alteration genetique deliberee chez les animaux et en particulier chez l'homme».* De telles règles éthiques fondamentales doivent rester présentes à l'esprit de chacun et nous rappeler que toute découverte scientifique peut devenir le meilleur ou le pire des outils entre les mains de l'homme. Avec ceci à l'esprit, *«la connaissance scientifique ainsi acquise dans le domaine de la genomique, de la proteomique et de l'evolution moleculaire a aussi valeur philosophique et merite d'appartenir au patrimoine culturel de l'humanite».*

Pascale Piguet

Piguet P, Poindron P (eds): Genetically Modified Organisms and Genetic Engineering in Research and Therapy. BioValley Monogr. Basel, Karger, 2012, vol 3, pp 1–32

Genetically Modified Organisms: Concepts and Methods

Pascale Piguet[a] · Philippe Poindron[b]

[a]Pharmazentrum, University of Basel, Basel, Switzerland; [b]Université de Strasbourg, Strasbourg, France

Abstract

Genetically modified organisms (GMO) raise societal, political and ethical concerns. They inspire strong resistance or, conversely, enthusiastic assent. What problems are faced in the field of genetic modification, and what scientific questions does it contribute to answer? The aim of this volume is to give an overview of genetic engineering from the technical aspects of transgenesis to their applications in research. Be it through the use of individual transfected cells or GMO, these applications cover a broad array ranging from disease-oriented research (but not only) to important perspectives for therapy. Gene therapy is fully described in full-length articles devoted to the use of this technique in cancer and in muscular disease. In this section, we shall define the following concepts: organisms, modification of germen and modification of soma, vectors and vector of expression, and transgene. We will then briefly describe the methods currently used to create GMO. A short description of the main transgenic organisms will be given in the first chapter. We do not intend to be exhaustive, but demonstrative and to give our readers objective elements for building their own opinion. Genetically modified plants are not covered in this chapter.

Defining and Illustrating Basic Concepts

Organism

In his introductory lecture to the International Congress for Virology held in Strasbourg, in August 1981, André Lwoff brilliantly defined what an organism is. For this purpose, he compared the pertinent properties of several biological entities (table 1). According to Lwoff [1], the main and specific properties of an organism are the capacity to evolve in an independent manner, contrary to cells which are, as organisms, living entities, but cannot evolve outside them. However, it should be mentioned that cultured cells, especially cell lines, or cell

Table 1. Respective properties of several biological entities

	Growth	Division
Protists, Fungi, Planta, and Animalia	+	+
Cells	+	+
Mitochondria, chloroplasts	+	+
Cytoplasmic membrane	+	+
Chromosomes, plasmids	0	0
Molecules, virus, and cellular organelles	0	0

Adapted from Lwoff [1].

strains can modify their properties all along the passages. However, they are not organisms since these alterations depend on experimental but not natural conditions.

It results from this definition that organisms, properly said, are biological entities belonging to all kingdoms of living creatures: animals, plants, fungi, protists.

Modification of Germen and Modification of Soma
An organism is able to multiply and reproduce itself. It results that, properly said, a GMO is an organism the germen of which has been altered either by insertion of one or several foreign genes originating from the same species or genus, from other more or less further-related organisms (transgenic organisms), or by deletion or inactivation of selected genes (knock-out (KO) organisms). We will focus this paper mainly on the first class of GMO, i.e. organisms modified by the insertion of new genes into their germinal genome.

It is also possible to modify the soma of an organism. The modification may theoretically affect all somatic cells, a given tissue, or only a specific cell type. This strategy is at the basis of gene therapy but has also been used for research purposes, i.e. to screen for specific genes – such as cancer genes – within a given tissue [2–4]. Since both types of modification (germen and soma) result in genetic modifications, we will briefly discuss the methods used in these two cases since they differ widely.

It is conceptually important not to confuse organisms with vectors, and especially viruses such as *Lentivirus* or *Adenovirus* that are frequently used for introducing gene sequences into target cells, as in therapy or transgenesis. Viruses are not organisms.

Gene Vectors or Gene Carriers for Gene Therapy
The name gene vector or gene carrier is often given to any entity that allows the introduction of genes into target tissues or cells. Numerous families of viruses

Piguet · Poindron

have been used as vectors of genes. Alternatively, synthetic molecules have been designed to allow the entry of genes into cells. Such vectors should also be non-toxic, poorly or nonimmunogenic; they may be natural or synthetic polymers or consist of nanoparticles designed for entering all cells or a specific class of cells of an organism. This class of vector is preferentially named gene carrier. It is associated with gene sequences and other elements that improve both the penetration and guidance of selected genes toward the nucleus. The conception of such artificial vectors resembles that of artificial viruses, without the adverse properties often associated to natural viruses that are immunogenicity and triggering of an immune response rapidly rendering them ineffective on reiterated administration.

Use of nude plasmids has also been proven efficient. In this situation, only the gene vector is used.

Vector of Expression
The name vector of expression rather applies to gene sequences especially designed to allow correct expression of a given gene in a given tissue, in all tissues, or in a zygote. In the case of gene therapy, this implies that regulating sequences of bases located upstream of the coding sequences be chosen correctly. The situation is different when the aim of genetic modification is to create a GMO.

Tools for Genetic Modification

The Case for Genetically Modified Organisms
Transgenic animals are essential research tools, whether they are used to address basic biological questions or to develop preclinical models of human diseases. In particular, transgenic mice have played a fundamental role in investigating tissue-specific gene expression, oncogenesis, and developmental mutations [5].

In order to successfully achieve transgenesis of animal eukaryotic cells (the case for prokaryotic organisms and plants will not be envisaged here), it is necessary to make the selected transgene (a) enter the recipient cell and (b) be integrated into its genome. These two steps have undergone numerous improvements during the past decade. Genetic methods of transgenesis can be divided into (a) transgene overexpression and (b) gene knockout techniques. Although their aims are not identical, both involve the two steps cited above.

The first animals to carry experimentally inserted genes were mice produced by injecting retroviral Simian virus 40 (SV40) DNA into the cavity of blastocysts [6]. Even though a large percentage of animals derived from the embryos carried SV40-specific DNA sequences in some of their tissues (mosaic integration), they neither incorporated the SV40 DNA into their germ cells nor expressed the viral genes. Germ line incorporation was later obtained by exposing mouse embryos to an infectious strain of Moloney leukemia virus. This first strain of transgenic mice (transmitting the virus as Mendelian genes) developed leukemia after birth, suggesting that viral genes were retained and expressed in adult animals [7]. Moreover, this demonstrated that embryos could harbor foreign DNA and still develop to term.

Most cell transfections involved a selection strategy to identify the cells into which the foreign DNA was successfully incorporated [8]. In a next step, the possibility of directly injecting DNA into mammalian cells was explored. In 1980, Gordon et al. [9] introduced foreign DNA into mouse embryos by injecting it into pronuclei, and Gordon and Ruddle [10] subsequently demonstrated the transmission of the introduced foreign DNA to offspring.

How to Make a Transgene Enter a Recipient Cell?
Theoretically, three kinds of cells can be chosen for achieving transgenesis: in vitro fertilized egg, embryonic stem cell and spermatozoa.

Original Method. The original method developed to generate transgenic organisms therefore involved the direct microinjection of foreign DNA. With nuclear injection of DNA, investigators rapidly reported germline transmission of transgenes, expression of transgenes [11], and a phenotype associated with transgene expression [12]. Pronuclear microinjection consists of injection of a small volume of fluid containing the gene of interest into a pronucleus of a zygote. Then, the zygotes are transferred to a foster mother. The generation of transgenic animals through the injection of naked plasmid DNA into the male pronucleus of fertilized oocytes has been standard practice for several decades. Nevertheless, the success of this technique has been largely limited to mice [5]. Even though pronuclei are clearly visible in many species, it is necessary in some other species to centrifuge embryos to displace the optically opaque cytoplasm in order to see the pronuclei [13]. After problems concerning the embryo viability were solved, several research groups have been able to produce transgenic rabbits, sheep and pigs by pronuclear microinjection [14, 15]. In swine, although the detection of transgene integration ahead of embryo transfer is possible, it is not practically convenient following pronuclear microinjection. For this reason, most of the piglets resulting from microinjection protocols are nontransgenic.

The generation of transgenic mice takes approximately a year from making the DNA construct to establishing a phenotypically characterized transgenic mouse strain [16]. For standard expression constructs, however, the classical pronuclear injection approach is still the major technique used in mice that consistently generates a single integration site. Because the standard approach of microinjection involves the random integration of DNA into the genome, it was considered nearly impossible to obtain embryos bearing targeted transgene integration by this approach. For a while, this feature restricted the technique to gain-of-function studies in the mouse [17]. At the end of the 1980s, disruption of genes in mice was developed by using cultures of pluripotent embryonic stem

(ES) cells [18, 19]. By maintaining their pluripotency, ES cells can regenerate all mouse tissues, including germ cells [20]. Transgenic mice are made by targeting a specific gene in ES cells, then introducing the modified cells into blastocysts, from which animals with the targeted gene can be obtained [21]. The application of gene targeting to these cells resulted in the creation of loss-of-function mutations in the mouse [22, 23].

The use of mouse embryonic stem cells to transfer a predetermined genetic modification to a whole animal is now routine [24]. However, gene targeting has not yet been achieved in mammals other than mice, due to the lack of functional embryonic stem cells in experimental protocols. An alternative procedure has therefore been developed which makes use of nuclear transfer from cultured somatic cells [25].

Nuclear Transfer Technology. Major progress has occurred since the first transgenic large animals – including pigs – were obtained through pronuclear microinjection. These advances led to the first knockout of a gene in a large animal model by combining genetic engineering on cultured somatic cells with nuclear transfer (somatic cell nuclear transfer, SCNT) [26].

Nuclear transfer was initially developed to address the possibility that differentiated cells retain all the genetic background necessary to generate an entire organism by cloning animals from differentiated cells [27]. This technique has allowed studying the role of genetic and epigenetic alterations during development and disease. The nuclear transfer procedure utilizes oocytes as the cytoplasm donor (cytoplast). The genetic material of the cytoplast is removed, after which a donor nucleus (karyoplast) is injected into the perivitelline space of the enucleated oocyte or injected into the cytoplasm of the oocyte [28]. After a fusion process, the oocyte is activated by either chemical or mechanical stimulation. Successful activation initiates development to the blastocyst stage, followed by transfer to foster mother. Live embryos have been reconstructed by SCNT using either electrofusion or piezoelectric microinjection (PEM) [29]. ElectroFusion consists of fusing a whole cell to the oocyte cytoplasm and avoids an extensive manipulation of the donor cell. On the other hand, PEM makes use of piezo pulses in order to break the donor cell membrane for penetrating the oocyte.

By performing the genetic engineering in vitro and transferring the manipulated genome to the enucleated oocytes, gene targeting has been made possible, both for knocking out a specific gene or for knocking in transgenes [30].

In pigs, nuclear transfer technology has replaced traditional microinjection. Modifying in vitro a somatic cell line by transfection, breeding and screening for transgene integration is no longer required. This has allowed expression in a single gestation period [31]. A limiting factor is, however, the difficulty in genetically modifying cells. Very low gene targeting efficiencies are achieved when using homologous recombination in somatic cells, as opposed to mouse ES cells [32, 33]. In order to improve such gene targeting efficiency, efforts have been made to decrease the number of random integration events by using promoter-trap strategies in homologous recombination systems. This, however, requires expression of the target gene in the nuclear donor cell [34]. Since fetal fibroblasts are among the few cell types adapted to nuclear transfer, the gene of interest must be expressed in the former for a successful result. Another limiting factor for using primary somatic cells is their relatively short life span in culture. The efficiency of nuclear transfer also decreases when donor cells are extensively cultured [35].

Even though SCNT works better in pigs than in other large animals [36, 37], SCNT has become the leading tool for obtaining animals from genetically engineered somatic cells.

Sperm-Mediated Gene Transfer. The first attempt to introduce exogenous DNA into intact spermatozoa was made 40 years ago by Brackett et al. [38]. Twenty years later, several authors claimed to have succeeded in introducing DNA into sea urchin spermatozoa [39, 40]. However, at that time, these results were contested [41] up to the discovery that they were quite valuable, the lack of reproducibility being probably imputable to the presence in variable amounts of an inhibitory factor (MW: 30–35 kDa), so-called IF-1, in the seminal fluid of either mammals or sea urchin. Finally, it was proven [42] that several DNA-binding proteins can regulate entry of DNA into spermatozoa and are differentially expressed by them, depending on their maturity. It is now accepted that under suitable conditions, spermatozoa can uptake foreign DNA, stable integration of the transgene occurring in about 10% of the progeny (for a detailed review, see [43]).

Three methods have been developed to introduce either linear or circular DNA into spermatozoa: (a) pretreatment of spermatozoa with DNA followed by assisted fertilization, (b) artificial intracytoplasmic sperm injection after sperm/ DNA interaction, and (c) restriction enzyme-mediated integration into nuclei of spermatozoa followed by their introduction into unfertilized eggs.

(a) Pretreatment of spermatozoa with foreign DNA.

To incorporate foreign DNA into the head of spermatozoa by simple contact seems the most direct way to achieve uptake of a putative transgene into male gametes. As previously mentioned, the success of the operation (transmission of the transgene to the offspring and expression of reporter gene) depends on multiple, uncontrolled factors. It is the reason why several techniques have been developed to improve the critical step of gene transfer into spermatozoa. In order to facilitate entry of exogenous DNA into spermatozoa, some authors submitted them to electroporation [44], without affecting their capacity to fertilize. This technique was used for transfecting spermatozoa of mollusks, fish (included zebrafish), pigs and cattle. In each case, a small proportion of the offspring expressed the transgene.

(b) Intracytoplasmic sperm injection after sperm/DNA interaction.

In order to overcome the death of spermatozoa consecutive to treatments aimed at helping the entry of DNA, alternate techniques have been developed.

Spermatozoa were incubated with DNA after having been submitted to several freeze-thaw or freeze-drying cycles, or permeabilization by Triton X-100, and then injected to oocytes. This technique has proved to be very efficient; it has been applied to mouse or monkey [45, 46].

(c) Microinjection of transfected sperm nuclei to oocyte.

It is also possible to directly transfer transfected nuclei of spermatozoa to the oocyte. In order to facilitate integration of the transgene into the genome of the recipient cell, the technique of restriction enzyme-mediated integration was used [47]. Such a treatment indeed resulted in decondensation of genomic DNA. This method was successfully applied to *Xenopus*, zebrafish and cattle. In order to facilitate integration of the transgene to DNA, lipofection was sometimes used. After being treated with exogenous DNA, sperm nuclei were injected into the oocytes.

Viral-Mediated Transgenesis. Viral vectors are very efficient in transferring genes, because the viruses from which they originate possess sophisticated machinery adapted to cell entry, transport and expression of their genomes in the nucleus of their hosts [48]. Several types of viruses have been used for transgenesis, such as lentiviruses, adenoviruses or adeno-associated viruses.

Lentiviruses have been successfully used into zygotes or early embryos from a wide variety of species including mice, rats, pigs, cows and chickens [49–53]. Lentiviruses are members of the Retroviridae family of viruses, which convert their RNA into DNA and integrate them into the genome of target cells [54–56]. Lentiviral vectors (LVs) present the interesting property of being able to stably transduce dividing and nondividing cells [57]. Transgene expression has been observed to spread over multiple generations. The phenomenon of silencing – a loss of transgene expression over time – has not been widely observed following lentiviral transgenesis [58]. In mice and rats, the rate of transgenesis using lentiviral transduction exceeds 50% [59]. In pigs, lentiviral gene transfer is extremely efficient, 80–100% of the offspring animals being transgenic after oocyte or embryo infection, or after somatic cell culture infection [60]. However, because of the risk associated to multiple integration of transgene into the genome, possibly leading to oncogene activation, insertional mutagenesis or silencing of viral sequences, and very frequently mosaicism in founder animals, this method cannot be used for producing transgenic animals for therapeutic use [61].

The typical procedure consists of injecting the lentivirus particle beneath the zona pellucida of a fertilized oocyte, which penetrates by fusion between the viral and plasma membranes. The viral RNA genome then undergoes reverse transcription, yielding a double-stranded DNA copy that integrates into the host cell chromosomes.

LVs have been made from several types of lentiviruses such as HIV-1 [62], HIV-2 [63], feline immunodeficiency virus [64], equine infectious anemia virus [65] and simian immunodeficiency virus [66]. The most commonly used is, however, the human immunodeficiency virus (HIV), whose molecular biology has been extensively investigated [67].

HIV-1-based lentiviral vectors include a transducing vector – used to transfer the gene(s) of interest to target cells – and separate helper plasmids. The separation into such components might decrease the risk of recombination leading to potentially harmful replication-competent lentiviruses. The first LVs were pseudotyped with the native HIV envelope protein, which limited their tropism to CD4-expressing cells, the natural target of HIV [48]. Since then, HIV-1-based lentiviral vectors have been pseudotyped with heterologous envelope proteins in order to confer broad tropism for transduction of a wide variety of mammalian cell types. HIV-1-based lentiviral vectors have undergone improvement [48, 68]. These have mainly concerned the vector and helper design constructed as plasmid DNAs to be transfected into cells to produce viral vector particles. In some cases, some heterologous elements (such as the woodchuck hepatitis virus posttranscriptional regulatory element) have been used to improve the properties of the transducing vector. Inducible expression has been created using reversible systems such as tetracycline trans-activators or repressors and irreversible systems such as short hairpin RNA (shRNA) expression using Cre–loxP recombination (to allow study of genes that would otherwise generate embryonic lethality) [69, 70].

The kinetics of lentivirus integration into the modified genome determines the degree of genotypic mosaicism in the generated animal. Integration events occurring after the first cell division will, indeed, limit the presence of individual proviruses to only a subset of cells therefore conditioning their rate of transmission to the progeny. Differences in the degree of mosaicism have been reported depending on the species: thus, transgenic mice have been reported to demonstrate much less mosaicism than transgenic rats [5]. Furthermore, the integration site will influence their potential to exert insertional mutagenesis (cis-acting effects on the host genome). Studies have shown that vectors derived from murine leukemia virus (MLV) tend to integrate in and around promoters, whereas the HIV-derived vectors rather target transcribed regions [71–73]. Noteworthy, when MLV vectors are used to infect early mouse embryos, they are rapidly silenced during development, in contrast to their lentiviral counterparts [71].

One of the limitations of transgenic animal models is that the transgene is present in every cell of the animal. Now, many disease processes are first limited to a few cells of a specific cell type. Mimicking such localized processes by restricting expression of a transgene to particular cell types may be an interesting strategy. This technology may be particularly useful for research applications or as a mechanism to increase vector safety. Thus, the combination of positively regulating tissue-specific promoters with negatively regulating microRNAs can induce very precise transgene expression in specific cells or tissues [74]. The induction of glioblastoma tumors was reported in adult immunocompetent mice by injecting a Cre-loxP-controlled lentiviral vector expressing the Ras and AKT oncogenes [75]. Thus, cancer research should highly profit from such technology as animal models for cancer can be made by directly injecting LV-expressing oncogenes into appropriate tissues of relevant species. Such animal models should provide insights into the pathogenesis of disease processes.

The use of LVs for genetically modifying primary cells has had a growing impact on research. Thus, the study of cell migration and cell function in animal models has benefited from in vivo imaging studies of transduced cells [48]. This was initially the case in mice with the in vivo imaging of hepatocytes transduced with the green fluorescent protein [76]. Imaging studies of in vivo cell distribution and transgene expression were then applied to various experimental models, including nonhuman primates [77].

Like the lentiviruses, recombinant adenovirus and recombinant adeno-associated virus display a broad host range and are able to infect proliferating as well as quiescent cells [78]. Adenoviruses are nonenveloped viruses with a linear double-stranded DNA genome which are associated with the common cold and cause respiratory, intestinal, and eye infections in humans [79]. In nature, wild-type Ads do not integrate their DNA into the host cell genome [80]. They generally replicate as linear, extra-chromosomal DNA elements in the nucleus [81]. There is evidence that recombinant Ads display a relatively low integration frequency: following transduction of eukaryotic cells, it is thought that adeno-viral vectors mainly persist as episomal DNA molecules. Because Ad genomes do not integrate into the host cell chromosome, this makes them safe vectors for mediating transient transgene expression. Various studies have shown that excision of DNA sequences with stabilizing genetic elements from the AdV genome can result in persistent transgene expression [78].

Apart from therapeutic applications, adenovirus-based vectors are used as a tool for efficient delivery of foreign DNA into the desired target cell. Engineered Ad vectors lack the early gene E1, the first transcriptional regulatory factor to be produced during the viral life cycle, which makes them replication-deficient. Thanks to a decrease in the number of viral genes present in the Ad vector, cyto-toxicity is diminished [82].

Most preclinical studies undergone to explore novel therapeutic approaches using replication-deficient recombinant adenovirus have been performed with Ad5 and Ad2 [80]. After systemic in vivo administration, Ad5 shows a natural tropism for liver: this was shown to result from binding of the blood coagulation factor X to the surface of the capsid and the hepatocyte, a predominant target cell [83]. In rodents, a single intravenous injection of high dose Ad has been shown to result in life-long phenotypic correction of genetic diseases [84, 85]. A period of transgene expression up to 964 days has been reported in baboons using an AdV for hepatic transduction [86].

Another class of frequently used viral vectors derives from the adeno-associated virus (AAV). AAV is a small nonenveloped, single-stranded DNA virus belonging to the Parvoviridae family [87–89]. Among the six AAV sero-types identified in primates, AAV type 2 isolated from humans has been the most extensively studied. AAV-derived vectors cross the cell membrane and deliver DNA molecules straight to the nucleus. A significant advantage of AAV vectors is that they can efficiently transduce most cell types and organs [24].

AAV vectors can be constructed using simple PCR-based methods, thus constituting an attractive alternative for gene targeting studies. In addition to expressing transgenes from randomly integrated or episomal vector genomes, AAV can be used for the targeted modification of homologous chromosomal sequences [90]. AAV single-strand DNA is highly efficient in integrating into the host genome through homologous recombination, even when performed in somatic cells [90, 91]. Thus, targeted modifications such as single basepair substitutions, deletions and insertions of up to 1.5 kb have been reported with AAV vectors [91–93]. In 2008, using AAV-mediated gene targeting and SCNT, Rogers et al. [34, 94] produced a CFTR-null (cystic fibrosis transmembrane con-ductance receptor-null) pig. Importantly, using the AAV-mediated approach for gene targeting was very efficient for knocking out the CFTR gene which is not expressed in fibroblasts.

Limitations of AAV-mediated gene targeting include the small packaging size of the vectors, the inability to target nondividing cells and a relative inefficiency of targeting silent loci. Moreover, the use of AAV-mediated gene targeting for therapeutic applications is limited by the random insertion of the vectors into the genome, which may result in harmful mutations [24]. However, gene tar-geting through recombinant adeno-associated virus (rAAV) was also shown to overcome limitations linked to the use of linear DNA fragments in homologous recombination (see below).

How to Facilitate Insertion of Transgene into a Recipient Genome?
Enzymatic Engineering. Several strategies have been developed to target inser-tion of transgene in selected portions of the genome. Transposons are mobile genetic elements. In the presence of so-called enzymes transposases, they can move from one part of the DNA to another. With these 'jumping genes', it is pos-sible to efficiently and accurately deliver selected segments of DNA in vertebrate cells. For this purpose, several transposons have been used, such as *Sleeping beauty, Passport Tol2* or *piggyBac*, in combination with their corresponding transposase expression construct. This method offers several advantages: it increases the efficiency of integration, precisely integrates a single copy of the transposon in one or several selected sites of the genome, and avoids the concat-enation of transgene which would result in a dramatic decrease in gene expres-sion (for instance, see [95]).

Another technique developed for establishing knockout and knockin strains of mice is that of the Cre Lox system (see, for instance, [96]). Briefly, this tech-nique makes use of Cre (*causes recombination*), a recombinase enzyme, produced by the phage P1. This enzyme is able to cut DNA sequences located between two lox P sites. LoxP sites are short DNA sequences that are recognized by Cre. The resulting DNA ends each have a half lox P site; these ends are then ligated by Cre. In its initial version, this technique made use of two strains of mice. One transgenic strain constitutively expresses Cre together with sequences needed for its expression (nonspecific or tissue-specific promoter). The other transgenic strain contains two lox P sites flanking the gene to be deleted associated to a drug selection marker. Crossing the two strains results in appearance of animals knocked out for the gene inserted between the lox P sites. More sophisticated versions of this technique allow producing so-called knocking strains of mice, harboring an allelic form of targeted gene.

In the frame of this section, a promising approach should also be mentioned, i.e. that of zinc finger nucleases which greatly improve the efficiently of gene targeting. These nucleases introduce double-stranded breaks in target genes therefore stimulating endogenous repair process and homologous recombination [97, 98].

SiRNA. Cells can be transfected with siRNA, which leads to degradation of targeted mRNAs by endonuclease, and to a dramatic decrease in the amount of protein encoded by these targets.

Additional Remarks. In some systems, for instance, lentiviral-mediated transgenesis, it has been proven that drugs such as cytokines or proteasome inhibitors can increase gene transfer into stem cells [99]. More specific techniques restricted to a defined class of organisms will be mentioned in the corresponding sections.

The Case for Gene Therapy
Viruses
Numerous families of viruses have been considered as potentially useful gene vectors. A virus, indeed, often possesses surface structures that specifically recognize specific structures of the cell membrane, so-called viral receptors. The specific recognition confers to the natural viruses' more or less tissue or cell specificity, rather widely exploited to restrict the delivery of a given gene to the desired cell or tissue target. In this limited space, it is not possible to exhaustively detail how the specific recognition of a cellular viral receptor has been exploited to genetically modify a cell. We shall limit illustration of this concept to a few examples.

Modifying the viral capsid may also be useful for a better or extended targeting. For example, deletion of adenoviral fiber knob and replacement with a biotin-accepting peptide allows the construction of a vector able to reach the desired cells or tissue, when combined with an antibody directed against a specific membrane structure expressed by the target cells [100].

Among the most common viral vectors developed are adenoviruses [101] (expression of RNAi activators) and lentiviruses [102]. But other families of viruses have also been studied for their capacities to be used as pertinent vectors, adeno-associated viruses [103], herpesviruses (treatment of cancer [104], often used under a modified form [105]), foamy viruses [106], and coxsackieviruses [107]. The viruses can be used both for gene therapy and transgenesis (see above).

Nonviral Vectors

Despite the high gene transfer efficiency of viral vectors, the induction of a host inflammatory and immune response limits their use in gene therapy [108]. For this reason, researchers have tried to develop nonviral carriers such as the most investigated cationic polymers and cationic liposomes. These chemical vehicles form electrostatic complexes with pDNA defined, respectively, as 'polyplexes' and 'lipoplexes' [109]. Nanoparticles are formed from the combination of cationic liposomes or polymers with nucleic acids in buffered solution [110]. These particles adsorb to the cell surface by electrostatic interactions between the positively charged complexes and the negatively charged cell surface [111–113], and enter cells by endocytosis or endocytosis-like mechanisms. The condensation of DNA with polycations or cationic lipids protects it from degradation by nucleases [114, 115].

Natural Biopolymers. Although a majority of nonviral gene delivery systems are based on synthetic polymers, natural polymers offer distinct advantages and are generally nontoxic, biocompatible and biodegradable. Some natural polymers studied for gene therapy include collagen, gelatin, chitosan, alginate, and their modified derivatives [116]. Chitosan is one of the most reported naturally derived gene carriers. Obtained from the partial deacetylation of chitin (a natural polysaccharide from crustacean shells), chitosan has low immunogenicity, strong affinity for DNA and can form microspheric particles [117]. Chitosan/DNA complexes have been reported to efficiently transfect different cell types such as B16 melanoma cells [118], African green monkey kidney cells (COS-1), human epithelial colorectal adenocarcinoma cells (Caco-2) [119], human embryonic kidney cells (HEK293), human osteosarcoma cell line (MG63) [120], human epithelial cervical cancer cells [121, 122] and primary chondrocytes [123]. Factors such as the degree of acetylation, molecular weight and charge of chitosan, plasmid concentration, charge ratio of amine (chitosan) to phosphate [124], serum concentration, pH of the media, and cell type have been shown to affect the efficiency of transfection [116, 117, 125]. Thanks to their mucoadhesive properties, chitosan-based delivery systems have been successfully applied to oral and nasal gene therapy systems. In 1999, Roy et al. [126] demonstrated an immunologic protection against peanut allergy in mice that were orally administered chitosan/DNA nanoparticles containing a dominant peanut allergen. In a BALB/c murine model of allergic asthma, intranasal administration of chitosan-IFN-γ nanoparticles increased the production of IFN-γ whereas epithelial inflammation was inhibited within 6 h of delivery [127]. Chitosan nanoparticles have also been used as a carrier for delivering factor VII DNA to hemophilia mice [128]. In 2009, an immune response against allergy was induced in BALB/c mice after oral delivery of chitosan/DNA nanoparticles encoding the Der p2 mite allergen [129].

Collagen is another natural polymer that has been extensively studied for gene transfer applications. In a comparative study, the transfection levels obtained between native and methylated collagen complexed with the luciferase encoding plasmid pRELuc were analyzed in vitro and in vivo following intramuscular injection [130]. Interestingly, methylated collagen/DNA particles exhibited higher transfection levels than native collagen/DNA particles whereas the opposite was true in vivo. These data suggest that further progress is necessary to understand the mechanisms of gene delivery before being able to fully exploit the benefits of collagen transfection systems.

Another natural polymer, gelatin, has been shown to be a potential viable gene delivery tool. In a cystic fibrosis model, in vitro transfection of human bronchial epithelial cells defective in CFTR-mediated transport, using a gelatin/pCFTR (cystic fibrosis transport regulator encoding plasmid) resulted in functional recovery of transport [131]. In 2010, cationized gelatin nanoparticles loaded with immunostimulatory RNA oligonucleotides led to an efficient antitumoral immune response in two different mouse tumor models [132].

Artificial Polymers. The use of cationic polymers for gene transfer was introduced by Wu and Wu [133] who complexed a plasmid to an asialoorosomucoidpoly-l-lysine conjugate and demonstrated some chloramphenicol acetyltransferase activity as a measure of gene transformation in hepatoma cells.

Polyethylenimine (PEI), the most common polymer used for DNA delivery, consists of multiple ethylenamine units and is water soluble [134]. In vitro, the gene expression levels of PEI have been shown to be comparable to those of viral vectors. Unfortunately, the predictive value of in vitro cell culture for in vivo gene transfer is low [135]. PEI-based DNA polymers were recently tested in humans for the treatment of bladder cancer. An extensive tumor regression was observed after intravesical vector application in treated patients [136]. One major drawback of synthetic vectors is, however, their toxicity profile: due to their high positive charge, an aggregation and reaction with blood components and erythrocytes has been reported in vivo [137]. Moreover, PEI cannot be efficiently metabolized, further potentially increasing its toxicity. The limited degradability of PEI precludes repeated systemic therapeutic administration, rather favoring a few localized applications. Numerous reports have suggested the use of analogous biodegradable polymers, which lead to smaller metabolites of lower toxicity. Of these, polyaminoesters have received much attention due to their interesting degradation profile, delivery efficiency and easy synthesis [135].

Nanoparticles and Related Carriers (Liposomes). Among the nonviral vectors, liposomes are particularly interesting for gene delivery applications and have a long history as gene delivery carriers [138]. In 1987, using the liposome-forming synthetic cationic lipids (N-[1-(2, 3-dioleyloxy) propyl]-N,N,N-trimethylammonium chloride; DOTMA), Felgner et al. [139] described a successful DNA-transfection protocol, which they called lipofection. Many efforts have been made over the last decades toward developing safe and efficient cationic lipids for use in gene delivery [140]. Lipofectamine, DOTAP, Geneporter, DOTMA – to name a few – have been widely used and commercialized.

Lipoplexes can be made by the interaction of a large variety of liposomes and DNA. The liposomes contain at least two parts: a cationic and a neutral lipid, sometimes named helper lipid [109]. Cationic lipids are amphiphilic molecules containing a positively charged polar head linked to a hydrophobic domain via a connector. Cholesterol and DOPE are often used as neutral lipids [109].

The transfection efficiency is not determined by one unique part of the cationic lipid structure but by a combination of them [115]. Various modifications have been performed at the hydrophobic parts, connector regions and head groups in order to improve DNA delivery to various mammalian cells [115].

In a recent study, Huang et al. [140] reported that cyclen-based cationic lipids are highly efficient in delivering genes towards several tumor cell lines, reaching a transfection efficiency level 5 times higher than lipofectamine. Using, in a mouse model of experimental autoimmune encephalitis, a single intrathecal injection of DNA coding for an immunomodulatory fusion protein, OX40-TRAIL, Yellayi et al. [141] demonstrated a decrease in the severity of the disease.

Despite significant advances in the field of gene therapy, the transfer of genes to target cells is still challenging. Nonviral vectors may overcome many of the problems encountered with viral carriers and might represent in the coming years a powerful alternative to viral-mediated therapy.

Genus or Species Used for Genetic Modification

Various kinds of organisms have been genetically modified. They belong to all kingdoms of living creatures: protists (especially bacteria), worms, insects, mammalians, plants (for a detailed classification of living creatures, see [142]).

Protists
Bacteria. Bacteria were modified either to produce molecules of therapeutic interest such as insulin or interferon or to treat inflammatory bowel diseases (colitis or cancer after being modified by insertion of a genetic sequence coding for the product of interest).

Lactobacillus lactis is a good candidate for treating inflammatory bowel diseases. It has been shown that replacement of Thy A gene of *L. lactis* with the coding sequence of mature IL-10 fused to the *L. lactis* (amino-terminal extremity) sequence coding for the secretion signal resulted in a genetically modified strain able to produce and secrete IL-10 at a good level [143]. These modified strains were successfully tested in mice displaying colitis [144]. Moreover, by rendering the strains dependent on the presence of thymine or thymidine, it was possible to control and/or limit expansion of the living bacteria, which ensures the safety of this kind of therapy [145].

Several genetically modified bacterial species were used to treat cancer (reviewed in [146]). Some attempts made use of genetically modified anaerobic bacteria such as *Clostridium novyi, Lactobacillus* sp. or *Bifidobacterium* to treat experimental tumors of animal. Since the toxin produced by *C. novyi* is extremely toxic, a strain deleted for the toxin coding gene was created. It accumulates in poorly oxygenated or anaerobic internal part of tumors. Due to the natural antitumoral properties of *C. novyi*, regression or size decrease of colonized regions was observed together with a non-negligible residual toxicity. Genetically engineered *Salmonella typhimurium* was also used in cancer therapy. The VNP209009 strain of *S. typhimurium* was deleted for *msbB* and *purI*. These alterations gave rise to an attenuated strain, strictly dependent on purine for its multiplication. It could multiply in tumor where purine is available, but not in spleen or liver which do not offer a convenient environment for its growth and multiplication. This vector showed long-lasting efficiency against a variety of experimental tumors, including metastases [147]. In addition, *S. typhimurium* can multiply both in aerobic and anaerobic conditions so that all parts of a given tumor can be colonized and regress. Such bacterial vectors can be further modified so as to produce molecules of therapeutic interest.

Since the topic has been widely treated elsewhere, we do not detail here the use of transgenic bacteria for the production of therapeutic peptides or proteins, such as insulin or interferon.

Animals
Worms: the Case for *Caenorhabditis elegans*, Rhabditidae
Impetus to *C. elegans* genetic modification was given by Brenner. In the mid-1960s, after having obtained brilliant successes using genetics and biochemistry for solving questions related to Bacteria and phages, he thought that time had come to investigate more complex systems, having, however, retained a modest level of complexity [148]. In 1974, in a famous paper, he described methods for isolating and mapping genes of mutagenized *C. elegans* [148]. This worm possesses many characteristics which rendered it a good candidate for studying molecular and genetic aspects of development, tissue organization, physiological apoptosis, and pathogenesis of inherited disease. He especially underlined the following features: (a) this worm has a small size (less than1 mm in length for adults individuals), is easy to propagate, rear and manipulate and can be stored on the long term after being frozen in the presence of glycerol; (b) he is transparent all along its life and therefore easy to visualize, at the very cellular level using conventional or sophisticated microscopy techniques; (c) it has a short

generation cycle (less than 3 days), and short life-span (about 3 weeks); (d) it displays a particular form of sexual dimorphism since it exists as self-fertilizing hermaphrodites and cross-fertilizing males; it is therefore possible to generate a very large-sized F1 (between about 300 and 1,000 individuals); (e) there exists a complete and invariant cell lineage map, from zygote to adult, both for the 959 somatic cells of the adult hermaphrodite and 1,031 somatic cells of the adult male; (f) the organ-base physiology of a nerve ring and nerve cords, intestine, muscle, hypoderm, secretory-excretory systems, gonads and uterus is entirely known; (g) the complete connectivity map for the 302 (adult hermaphrodite) and 391 (adult male) neurons is entirely established; (h) behavior of the worm (mechano-sensitivity, chimio-sensitivity, reaction to noxious stimuli, thermotaxis) is entirely elucidated; (i) techniques for transgenesis have been developed and well mastered and it is possible to flow sort individuals based on size and fluorescence; (j) the genomic sequence (about 19,700 coding sequences and 1,300 noncoding RNAs) is entirely deciphered; (k) there exist repositories for genomic and cDNA clones; (l) there exists a stock center for mutants, including null mutants for every gene, and (m) finally it is possible to consult free online resources (WormBook) and data bases (from [149]).

Using *C. elegans* as GMO, it has been possible to elucidate the three cell death pathways leading to cell apoptosis or necrosis, that is (a) the classical core apoptosis pathway, in which upstream regulators block activation of the caspase; (b) the neurodegenerin pathway which takes place in mechanosensory neurons and more evidently in intestinal cells, and is activated by mutations eliciting excitotoxic injury; this necrosis pathway is triggered by hypotonic, thermal, oxidative or hypoxic stresses (see [150, 151]) and (c) the serpin pathway, the protective element of which is an intracellular serpin (SRP 6) protecting lysosomes from stress-induced injury and which is very probably conserved in higher eukaryotes, including humans [152].

C. elegans has also been used for modeling human disease. Although, first considered as not really interesting for elucidating molecular mechanisms of inherited or acquired human diseases, the model has rapidly been proved to be heuristic and fruitful. For example, it has been possible to drive expression of Aβ1–42 human amyloid peptide in body wall muscles or neurons. Aβ1–42 inclusions appeared in both cell types in young adults, but only body wall muscle expression led to progressive paralysis of animals [153]. Interestingly, it has been possible to identify several proteins interacting with Aβ1–42, including heat-shock proteins (HSP) belonging to HSP-70 and HSP-16 families which play the role of chaperones. These studies also proved that cellular dysfunction can precede accumulation of Aβ1–42 inclusion bodies. Numerous other neurodegenerative diseases involve genomic amplification of CAG (Huntington's disease, spinal and bulbar muscular atrophy, dentatorubral-pallidoluysian atrophy, spinal-cerebellar ataxias types 1, 2, 3, 6, 7 and 17). Some of them were studied using *C. elegans* mutants. CAG repeats are translated into polyglutamine tracts.

Above a critical threshold (35–40 repeats), these polyglutamine tracts result in appearance of toxic phenotypes, with intranuclear or intracytoplasmic inclusions, neuronal dysfunction and cell death. These studies showed that polyglutamine repeats impair normal protein foldings and protein folding quality control systems [154]. Finally, using several *C. elegans* mutants, it has been possible to study different forms of inherited amyotrophic lateral sclerosis (by using transgenes of several forms of mutated superoxyde dismutase) or Parkinson's disease (α-synuclein mutant). Such an approach would have been impossible using mice.

Finally, it should been also mentioned that *C. elegans* has been used for studying host-pathogen interactions and as model of parasitic nematode. It is killed by numerous human Gram-positive or Gram-negative pathogens, and fungi. This property has rendered possible the molecular dissection of host-pathogen interactions. Although free-living, *C. elegans* shares with obligatorily parasitic human nematodes, genomic, anatomical, developmental, physiological and pharmacological properties. Therefore, it was largely used for screening of anthelminthic, and resistance to these drugs. Mutants of *C. elegans* in the three β-tubulin genes have been developed to elucidate mechanism of resistance to benzimidazoles, which are the major anthelminthic drugs.

Last but not least, due to its small size, *C. elegans* can be used for high throughput screening of drugs [155].

Insects
Silkworm Bombyx mori, *Bombycidae.* For centuries, the silkworm *B. mori* has been considered as an essential bioresource in Japan and China [156]. Today numerous mutant strains of this Lepidoptera are selected for useful characters, such as resistance to viruses, or color of silk. But, for a long time, it has been difficult to achieve transgenesis in *B. mori*. This was rendered possible by using a sophisticated system of transposon-based vectors. The *PiggyBac* transposable element is the most promising vector for transformation of blastoderm insect embryo. First isolated from a Lepidoptera, *Trichoplusia ni*, the cabbage looper, it was shown to be functional and efficient to also transform Diptera and Coleoptera [157], different to other insect transposons, the efficiency of which seems restricted to a narrower range of hosts. However, the method is complex and requires skilled operators. Indeed, it is necessary to separate the transposase gene and the transposon backbone, rendering the transposon unable to achieve autonomous mobility. Two different vectors of expression are simultaneously injected into the preblastoderm embryo: the plasmid coding for the nonautonomous transposon and the helper plasmid coding for transposase. The enzyme excises the transposon backbone and mediates the insertion of sequence of interest into the host genome.

Such a strategy has rendered possible, for instance, the functional analysis of peptidergic cells in the silkworm [157].

Anopheles sp., Culicidae, Anophelinae. Malaria is probably the most widespread parasitic disease affecting humans in sub-Saharan Africa. Because of the appearance of *Plasmodium* strains resistant to conventional antimalarial treatment, attempts have been made to find alternative methods for controlling the expansion of malaria. Among them, limitation of vector proliferation by genetic engineering has retained the attention of researchers [158].

One approach used to reach this aim was to identify the receptors for the proteins required for the parasite to cross the gut of *Anopheles* vector and invade the hemocoel after being ingested, and then to genetically engineer mosquitoes by driving them to produce proteins saturating these receptors sites, in order to block multiplication and subsequent transmission of the parasite.

It is well known that among the *Anopheles* genus (about 50 different species), there exist individuals that display natural refractoriness to *Plasmodium* and are unable to allow their multiplication. Identifying the involved gene would allow altering functions linked to sensitiveness, thus limiting the extension of *Plasmodium* transmission. However, the construction of a refractory construct is an uneasy task. The aim of such attempts is to replace populations of susceptible mosquitoes with populations of refractory mosquitoes. For this purpose, the use of transposable elements was first tested [159]. The elements are able to replicate in a host genome and are inherited by the progeny. Attaching a transposable element to a gene conferring resistance to *Plasmodium* certainly seems to be an elegant method. However, some difficulties arose, limiting the use of such approach: (a) for instance, the transposable element cannot be introduced into the genome of *Anopheles gambiae*, the principal vector of malaria in Africa; (b) the transposable element mutates and rapidly becomes inactive, and (c) increase in size by linking driven sequences leads to a declining activity. Obviously, further improvement is needed before any practical use can be taken.

Another promising approach consists of using the artificial sequence *Medea* [160], based on the discovery of a natural element in *Tribolium castaneum*, a species of flour beetle. *Medea* rapidly spreads through a population of mosquitoes and gives birth to offspring able to replace a population of mosquitoes which do not express it. *Medea* acts by killing the offspring of heterozygous females that do not inherit the allele. This name makes reference to Medea, a mythical magician who avenged on Jason by killing his children. The *Medea*-bearing population increases with generation. By attaching a gene of refractoriness to this element, it is hoped that resistant populations of *Anopheles* will replace the wild, susceptible populations of mosquitoes involved in malaria transmission.

Some authors also attempted to introduce into the genome of *Anopheles* a lethal dominant gene placed under the control of a female specific promoter, namely that of vitellogenin [161]. The expression of such a lethal gene is inactivated by treatment with tetracyclin. Briefly, the tetracyclin-repressible transactivator (tTA) protein is placed under the control of a promoter which determines the sex and developmental specificity of the system. When tTA is expressed, its binds to a specific DNA sequence tRE linked to an adjacent proximal lethal gene, the expression of which is controlled by tRE. In the absence of transactivator expression, i.e. in presence of tetracycline, the expression of the lethal gene is repressed. This allows maintaining colonies of genetically modified mosquitoes. In the absence of tetracycline, tTA is expressed, links to tRE and activates expression of the lethal gene, causing death of females. Males homozygous for the lethal gene, by mating with wild females, give birth to a heterozygous progeny, of which only males survive, resulting in a progressive disappearance of mosquito populations.

Fish: The Case for Zebrafish, *Danio rerio*, Cyprinidae

The zebrafish was first used to study vertebrate organogenesis [162]. It offered obvious advantages to researchers: the larvae are very small (few millimeters long), they are transparent; embryogenesis occurs ex vivo, and is quite complete within the 3 days following fertilization; a clutch of a pair of fish consists of approximately 100–300 embryos, and, last but not least, maintenance of a zebrafish colony is much cheaper than that of a mice colony (about 1,000-fold lower). All these characteristics render the zebrafish as a very attractive tool for observing the effects of genetic engineering on the phenotype.

The sequence of the zebrafish genome is almost completed. It is very easy to achieve genetic manipulations in a large number of individuals. Since embryogenesis occurs ex vivo, larvae are easily amenable to microinjection techniques. It is why several transgenic lines of zebrafish have been generated after microinjection of plasmidic or bacterial artificial chromosomes. Other methods made use of transposons, namely *Tol2* and *Sleeping Beauty*, a Tc1/mariner family of transposable elements [163]. At present, it has not been possible to generate zebrafish lines knockout for specific genes, because stem cell cultures cannot contribute to germ cell lineage. However, despite these rare limitations, the zebrafish model is one of the most interesting systems for studying fundamental questions concerning development, cancerogenesis [164], physiological functions and pathological processes (arrhythmia, demyelination/remyelination) [165, 166].

Amphibians: *Xenopus*, Pipidae

The genus *Xenopus* contains about 20 species. Two of them are especially known by the biologists: *X. laevis* and *X. tropicalis*. The first has been used for studying the mechanisms of early development, and the second as a model for transgenesis. Indeed, all species of *Xenopus* except one, *X. tropicalis*, are polyploid. This feature renders very difficult the attempt to produce genomic modifications and study their transmission to offspring. In addition, *X. laevis* has a generation time of 1–2 years, whereas *X. tropicalis* has a generation time of 6–9 months. These two properties confer on *X. tropicalis* a definitive advantage for production of transgenic strains.

Several procedures have been devised for this purpose [167]: (a) DNA microinjection in fertilized eggs [168] which, in case DNA persists in adulthood, often results in the appearance of a mosaic pattern of distribution; (b) germinal transgenesis using the restriction enzyme-mediated incorporation of linear DNA into the genome before the first cell division occurs [169]; (c) sperm-mediated gene transfer [170]; (d) use of transposable element [171]; (e) use of I-SCe I meganuclease, an enzyme isolated from *Saccharomyces cerevisiae* which recognize an 18 bp sequence [172]; (f) use of phi-C31-integrase [173] (see 'Methods' for details).

The sequence of the *X. tropicalis* genome is now completed [174, 175]. This offers scientists numerous opportunities for carrying out functional studies. Till now, they had focused their attention on retina and muscle development, together with metamorphosis. Fluorescent peptides have been used to determine the intracellular localization and the development fate of a given molecule in living cells [176]. They have also been used to track signaling networks with the aid of so-called transgenic sensor lines [177]. Several attempts have been made to select transgenic strains expressing nonfluorescent transgene products. The procedure is based on the introduction of *aphA-2* gene into the genome; this gene confers resistance to G418, a toxic antibiotic [178]. Other strategies are currently developed to perform selected cell ablation.

Mammalians

Mouse: Mus musculus, *Muridae, Murinae.* Mice are easily amenable to genetic modifications. This is why transgenesis has been extensively studied in, and applied to, these animals, particularly, but not solely, with the aim of establishing animal models of human diseases [96]. Humans and mice express about 30,000 genes, only 1% of which (about 300) are unique to either species [122]. This means that humans and mice share genomic sequence complexity. Inbred and congenic strains of mice can be easily established, therefore eliminating genetic variations, unique gene polymorphism and epigenetic influences depending on these differences. Mice have a short generation time and give rise to large litters. This character allows studying the effects of transgenesis on statistically significant populations. In addition, conditions of housing and experimental protocols can be standardized.

At difference with ES cells originating from nonrodent mammalians species or other vertebrates, mice ES cell lines can contribute to germ cell lineage. In the frame of this short review, it is not possible to describe all the applications of mouse transgenesis. We will only cite two studies illustrating interest and possible limitations of using mice for transgenesis.

In an extensive description of mice knocked out for one or two subtypes of nicotinic acetylcholine receptors subunits, Fowler et al. [179] extensively analyzed the behavioral effects resulting from such modifications. The aim was to identify the type of nicotinic acetylcholine subunit receptor putatively involved in tobacco addiction, or psychiatric disorders (anxiety, depression, schizophrenia). It was possible to appraise the contribution of discrete subunits involved in the nicotine dependence process (namely β_2).

Conversely, because of physiological differences between human and mice pancreas and lung physiology, none of the eleven strains of genetically modified mice developed to establish a relevant model of cystic fibrosis has reproduced the clinical signs of human disease [see Yan, Sun et al., 2009]. It is the reason why ferrets or pigs are preferred because of their similarities with humans in lung physiology.

When one desires to create a relevant model of a human disease using knock-out mice, it is therefore necessary to ascertain whether physiological similarities between mice and humans are sufficient.

Pig: Sus scrofa domesticus, *Suidae, Suinae.* Pigs have breeding characteristics which confer suitable properties for transgenesis studies [61]. They have a short duration of gestation and give birth to very large litters. Their oocytes are easily amenable to experimental manipulations. Efficient techniques of artificial insemination allow easy planning of experiments. It is the reason why both transgenic and KO strains have been generated. For purposes of organ transplantation, generating KO animals was an important challenge.

Among mammalians, the pig is indeed considered as the most suitable potential donor species of organs for solid organ transplantation in humans [31]. It resembles humans in its metabolic features and in contrast to the monkey, it is widely distributed throughout the world (and is not an internationally protected species). However, till now, stem cell cultures of pig are not available. Only transfer of genetically modified embryo to surrogate mother, sperm-mediated gene transfer, viral transgeneis or nuclear transfer technology to oocyte and transfer to foster mother are usable [61, 180].

Thanks to the genetic modification of donor pigs, the problem of hyperacute rejection of the graft has now been solved. Pigs expressing CD55, CD46 or CD59 (human complement regulatory factors) have been created. Such modifications have been shown to suppress hyperacute rejection of heart, kidney or liver in a pig-to-primate system, with a graft surviving for days to weeks before being rejected [180–185]. Problems linked to coagulation disorders, ischemia-reperfusion injury, innate immune response, rejection due to production of antibodies, and cell-mediated rejection are currently under intensive investigation.

Due to introduction of nuclear transfer technology, it is now possible to bypass the traditional and fastidious techniques of microinjection by using modification of cultured somatic cells, the nuclei of which are then introduced into enucleated of pig ovocyte before transfer into the foster mother (see 'Tools for Genetic Modification').

Primates

Because of their evolutionary relationship, nonhuman primates are endowed with numerous characteristics analogous to those of humans, especially genetic, neuro-anatomic, cognitive, behavioral and physiological properties (blood pressure, heart rate, menstrual cycle, total body oxygen consumption) as quoted by Kuang et al. [186]. These analogies render the monkey a very suitable model for studying human physiology, human gene function, and mechanisms of human inherited diseases together with preclinical research of novel therapies for the treatment of human diseases. For this purpose, numerous species have been used, originating either from the Old World or New World: chimpanzee, rhesus, cynomolgus, marmosets, squirrels monkey.

The first transgenic monkey *(Macaca mulatta)* was generated in 2001 [46]. More recently, Chan and his colleagues created monkeys which express a polyglutamine-expanded human huntington [187]. These animals presented numerous signs of Huntington disease, such as nuclear inclusions and neuropil aggregates in the brain. It has been possible to transmit green fluorescent protein transgene in the germline of the common marmoset *(Callithrix jacchus)* [188, 189], giving the possibility to use very small animals, with a relatively short cycle of reproduction, for creating several models of human diseases. However, marmosets are less closely related to humans than monkeys of the Old World, and this probably introduces limitation in generalizing the use of this species.

As pointed out by Kuang et al. [186], the use of monkeys as transgenic models is limited by the low efficiency of live birth baby monkeys. The most common method for obtaining transgenic monkeys makes use of lentiviral vectors which offer interesting characteristics and especially low toxicity, moderate insert size, and availability. However, numerous copies of lentiviral vectors are integrated which complicates the use of the transgenic individuals for breeding. They also mentioned that the cost of these techniques (sustained and large-scale production of transgenic monkeys) is prohibitive.

Those who are interested in producing transgenic *Macaca mulatta* will read with profit the detailed review by Chan and Yang [190].

Conclusions

After this incomplete survey of GMO, the time has come to conclude. We do not intend to give a critical appraisal or an ethical judgment on the techniques leading to the generation of GMO or on the use of GMO. Other authors in this monograph give their point of view on these very important questions. We prefer to summarize the current opinion of the vulgum pecus, using, however, an apologue to express it humoristically, that of Aesop's tongues. Let us cite a passage of Aesop's fables, published on-line by Forgotten Books: 'Xanthus invited a large company to diner, and Aesop was ordered to furnish the feast with the choicest dainties that money could procure. The first course consisted of tongue, cooked in different ways, and served with appropriate sauces. This gave rise to a deal of mirth and witty remarks among the assembled guests.

The second course, however, like the first, was also nothing but tongue, and so the third, and the fourth. The matter seemed to all to have gone beyond a jest, and Xanthus demanded of Aesop: "Did I not tell you, sirrah, to provide the choicest dainties that money could procure?" "And what excels the tongues?" replied Aesop. "It is the great channel of learning and philosophy. By this noble organ addressed and eulogies are made, and commerce, contracts, and marriages completely established. Nothing is equal to the Tongue." The company applauded Aesop's wit, and good-humor was restored. "Well", said Xanthus to the guests, "pray do me the favor of dining with me again tomorrow. And if this is your best", continued he, turning to Aesop, "pray, tomorrow, let us have some of the worst meat you can find". The next day, when dinnertime came, the guests were assembled. Great was their astonishment, and great the anger of Xanthus, at finding that again nothing but tongue was put upon the table. "How, sir" said Xanthus, "should tongues be the best of meat one day and the worst another?" "What", replied Aesop, "can be the worst that is not a part in? Treason, violence, injustice and fraud are debated, resolved upon, and communicated by the Tongue. It is the ruin of empires, cities and of private friendships". The company were more than ever struck by Aesop's ingenuity, and successfully interceded for him with his master.'

Mutatis mutandis, GMO resembles the Aesop's tongues. They are the best and possibly the worst scientific discovery, depending on their use. Consciousness, prudence, control and regulation certainly should render them not only acceptable but also useful for the health and welfare of humanity.

References

1   Lwoff A: Introduction au 5e Congrès International de Virologie, 2–7 août 1981, Strasbourg. Ann Virol (Inst Pasteur) 1981; 132:121–131.

2   Neil JC, Cameron ER: Retroviral insertion sites and cancer: fountain of all knowledge? Cancer Cell 2002;2:253–255.

3   Collier LS, Carlson CM, Ravimohan S, Dupuy AJ, Largaespada DA: Cancer gene discovery in solid tumours using transposon-based somatic mutagenesis in the mouse. Nature 2005;436:272–276.

4   Largaespada DA: Transposon mutagenesis in mice. Methods Mol Biol 2009;530:379–390.

5   Sauvain MO, Dorr AP, Stevenson B, Quazzola A, Naef F, Wiznerowicz M, Schutz F, Jongeneel V, Duboule D, Spitz F, Trono D: Genotypic features of lentivirus transgenic mice. J Virol 2008;82:7111–7119.

6   Jaenisch R, Mintz B: Simian virus 40 DNA sequences in DNA of healthy adult mice derived from preimplantation blastocysts injected with viral DNA. Proc Natl Acad Sci USA 1974;71:1250–1254.

7   Jaenisch R: Germ line integration and Mendelian transmission of the exogenous Moloney leukemia virus. Proc Natl Acad Sci USA 1976;73:1260–1264.

8   Graham FL, van der Eb AJ: Transformation of rat cells by DNA of human adenovirus 5. Virology 1973;54:536–539.

9   Gordon JW, Scangos GA, Plotkin DJ, Barbosa JA, Ruddle FH: Genetic transformation of mouse embryos by microinjection of purified DNA. Proc Natl Acad Sci USA 1980;77:7380–7384.

10 Gordon JW, Ruddle FH: Integration and stable germ line transmission of genes injected into mouse pronuclei. Science 1981;214:1244–1246.

11 Wagner EF, Stewart TA, Mintz B: The human beta-globin gene and a functional viral thymidine kinase gene in developing mice. Proc Natl Acad Sci USA 1981;78:5016–5020.

12 Palmiter RD, Brinster RL, Hammer RE, Trumbauer ME, Rosenfeld MG, Birnberg NC, Evans RM: Dramatic growth of mice that develop from eggs microinjected with metallothionein-growth hormone fusion genes. Nature 1982;300:611–615.

13 Wall RJ, Pursel VG, Hammer RE, Brinster RL: Development of porcine ova that were centrifuged to permit visualization of pronuclei and nuclei. Biol Reprod 1985;32:645–651.

14 Hammer RE, Pursel VG, Rexroad CE Jr, Wall RJ, Bolt DJ, Ebert KM, Palmiter RD, Brinster RL: Production of transgenic rabbits, sheep and pigs by microinjection. Nature 1985;315:680–683.

15 Al-Shawi R, Burke J, Jones CT, Simons JP, Bishop JO: A Mup promoter-thymidine kinase reporter gene shows relaxed tissue-specific expression and confers male sterility upon transgenic mice. Mol Cell Biol 1988;8:4821–4828.

16 Ittner LM, Gotz J: Pronuclear injection for the production of transgenic mice. Nat Protoc 2007;2:1206–1215.

17 Jasin M, Moynahan ME, Richardson C: Targeted transgenesis. Proc Natl Acad Sci USA 1996;93:8804–8808.

18 Evans MJ, Kaufman MH: Establishment in culture of pluripotential cells from mouse embryos. Nature 1981;292:154–156.

19 Martin GR: Isolation of a pluripotent cell line from early mouse embryos cultured in medium conditioned by teratocarcinoma stem cells. Proc Natl Acad Sci USA 1981;78:7634–7638.

20 Bradley A, Evans M, Kaufman MH, Robertson E: Formation of germ-line chimaeras from embryo-derived teratocarcinoma cell lines. Nature 1984;309:255–256.

21 Gossler A, Doetschman T, Korn R, Serfling E, Kemler R: Transgenesis by means of blastocyst-derived embryonic stem cell lines. Proc Natl Acad Sci USA 1986;83:9065–9069.

22 Thomas KR, Capecchi MR: Site-directed mutagenesis by gene targeting in mouse embryo-derived stem cells. Cell 1987;51:503–512.

23 Zijlstra M, Li E, Sajjadi F, Subramani S, Jaenisch R: Germ-line transmission of a disrupted beta 2-microglobulin gene produced by homologous recombination in embryonic stem cells. Nature 1989;342:435–438.

24 Sorrell DA, Kolb AF: Targeted modification of mammalian genomes. Biotechnol Adv 2005;23:431–469.

25 McCreath KJ, Howcroft J, Campbell KH, Colman A, Schnieke AE, Kind AJ: Production of gene-targeted sheep by nuclear transfer from cultured somatic cells. Nature 2000;405:1066–1069.

26 Wheeler MB, Walters EM: Transgenic technology and applications in swine. Theriogenology 2001;56:1345–1369.

27 Meissner A, Jaenisch R: Mammalian nuclear transfer. Dev Dyn 2006;235:2460–2469.

28 Onishi A, Iwamoto M, Akita T, Mikawa S, Takeda K, Awata T, Hanada H, Perry AC: Pig cloning by microinjection of fetal fibroblast nuclei. Science 2000;289:1188–1190.

29 Yu Y, Ding C, Wang E, Chen X, Li X, Zhao C, Fan Y, Wang L, Beaujean N, Zhou Q, Jouneau A, Ji W: Piezo-assisted nuclear transfer affects cloning efficiency and may cause apoptosis. Reproduction 2007;133:947–954.

30 Galli C, Perota A, Brunetti D, Lagutina I, Lazzari G, Lucchini F: Genetic engineering including superseding microinjection: new ways to make GM pigs. Xenotransplantation 2010;17:397–410.

31 Gock H, Nottle M, Lew AM, d'Apice AJ, Cowan P: Genetic modification of pigs for solid organ xenotransplantation. Transplant Rev (Orlando) 2011;25:9–20.

32 Sedivy JM, Dutriaux A: Gene targeting and somatic cell genetics–a rebirth or a coming of age? Trends Genet 1999;15:88–90.

33 Williams SH, Sahota V, Palmai-Pallag T, Tebbutt SJ, Walker J, Harris A: Evaluation of gene targeting by homologous recombination in ovine somatic cells. Mol Reprod Dev 2003;66:115–125.

Piguet · Poindron

34 Rogers CS, Hao Y, Rokhlina T, Samuel M, Stoltz DA, Li Y, Petroff E, Vermeer DW, Kabel AC, Yan Z, Spate L, Wax D, Murphy CN, Rieke A, Whitworth K, Linville ML, Korte SW, Engelhardt JF, Welsh MJ, Prather RS: Production of CFTR-null and CFTR-DeltaF508 heterozygous pigs by adeno-associated virus-mediated gene targeting and somatic cell nuclear transfer. J Clin Invest 2008;118:1571–1577.

35 Wilmut I, Beaujean N, de Sousa PA, Dinnyes A, King TJ, Paterson LA, Wells DN, Young LE: Somatic cell nuclear transfer. Nature 2002;419:583–586.

36 Lagutina I, Lazzari G, Duchi R, Turini P, Tessaro I, Brunetti D, Colleoni S, Crotti G, Galli C: Comparative aspects of somatic cell nuclear transfer with conventional and zona-free method in cattle, horse, pig and sheep. Theriogenology 2007;67:90–98.

37 Brunetti D, Perota A, Lagutina I, Colleoni S, Duchi R, Calabrese F, Seveso M, Cozzi E, Lazzari G, Lucchini F, Galli C: Transgene expression of green fluorescent protein and germ line transmission in cloned pigs derived from in vitro transfected adult fibro-blasts. Cloning Stem Cells 2008;10:409–419.

38 Brackett BG, Baranska W, Sawicki W, Koprowski H: Uptake of heterologous genome by mammalian spermatozoa and its transfer to ova through fertilization. Proc Natl Acad Sci USA 1971;68:353–357.

39 Arezzo F: Sea urchin sperm as a vector of foreign genetic information. Cell Biol Int Rep 1989;13:391–404.

40 Lavitrano M, Camaioni A, Fazio VM, Dolci S, Farace MG, Spadafora C: Sperm cells as vectors for introducing foreign DNA into eggs: genetic transformation of mice. Cell 1989;57:717–723.

41 Brinster RL, Sandgren EP, Behringer RR, Palmiter RD: No simple solution for making transgenic mice. Cell 1989;59:239–241.

42 Carballada R, Esponda P: Regulation of foreign DNA uptake by mouse spermatozoa. Exp Cell Res 2001;262:104–113.

43 Celebi C, Guillaudeux T, Auvray P, Vallet-Erdtmann V, Jegou B: The making of 'transgenic spermatozoa'. Biol Reprod 2003; 68:1477–1483.

44 Tsai HJ, Lai CH, Yang HS: Sperm as a carrier to introduce an exogenous DNA fragment into the oocyte of Japanese abalone (*Haliotis divorsicolor suportexta*). Transgenic Res 1997; 6:85–95.

45 Wakayama T, Perry AC, Zuccotti M, Johnson KR, Yanagimachi R: Full-term development of mice from enucleated oocytes injected with cumulus cell nuclei. Nature 1998;394: 369–374.

46 Chan AW, Chong KY, Martinovich C, Simerly C, Schatten G: Transgenic monkeys produced by retroviral gene transfer into mature oocytes. Science 2001;291:309–312.

47 Kroll KL, Amaya E: Transgenic *Xenopus* embryos from sperm nuclear transplanta-tions reveal FGF signaling requirements during gastrulation. Development 1996;122: 3173–3183.

48 Dropulic B: Lentiviral vectors: their molecu-lar design, safety, and use in laboratory and preclinical research. Hum Gene Ther 2011;22:649–657.

49 Lois C, Hong EJ, Pease S, Brown EJ, Baltimore D: Germline transmission and tissue-specific expression of transgenes deliv-ered by lentiviral vectors. Science 2002;295: 868–872.

50 Tiscornia G, Singer O, Ikawa M, Verma IM: A general method for gene knockdown in mice by using lentiviral vectors expressing small interfering RNA. Proc Natl Acad Sci USA 2003;100:1844–1848.

51 Fassler R: Lentiviral transgene vectors. EMBO Rep 2004;5:28–29.

52 van den Brandt J, Wang D, Kwon SH, Heinkelein M, Reichardt HM: Lentivirally generated eGFP-transgenic rats allow effi-cient cell tracking in vivo. Genesis 2004;39: 94–99.

53 Hofmann A, Kessler B, Ewerling S, Kabermann A, Brem G, Wolf E, Pfeifer A: Epigenetic regulation of lentiviral transgene vectors in a large animal model. Mol Ther 2006;13:59–66.

54 Naldini L: Lentiviruses as gene transfer agents for delivery to non-dividing cells. Curr Opin Biotechnol 1998;9:457–463.

55 Baltimore D, Verma IM, Drost S, Mason WS: Temperature-sensitive DNA polymerase from Rous sarcoma virus mutants. Cancer 1974;34(4 suppl):1395–1397.

56 Esposito D, Craigie R: HIV integrase structure and function. Adv Virus Res 1999;52: 319–333.

57 Naldini L, Blomer U, Gallay P, Ory D, Mulligan R, Gage FH, Verma IM, Trono D: In vivo gene delivery and stable transduction of nondividing cells by a lentiviral vector. Science 1996;272:263–267.

58 Dann CT, Alvarado AL, Hammer RE, Garbers DL: Heritable and stable gene knockdown in rats. Proc Natl Acad Sci USA 2006;103:11246–11251.

59 Dann CT: New technology for an old favorite: lentiviral transgenesis and RNAi in rats. Transgenic Res 2007;16:571–580.

60 Whitelaw CB, Radcliffe PA, Ritchie WA, Carlisle A, Ellard FM, Pena RN, Rowe J, Clark AJ, King TJ, Mitrophanous KA: Efficient generation of transgenic pigs using equine infectious anaemia virus (EIAV) derived vector. FEBS Lett 2004;571:233–236.

61 Sachs DH, Galli C: Genetic manipulation in pigs. Curr Opin Organ Transplant 2009;14:148–153.

62 Dull T, Zufferey R, Kelly M, Mandel RJ, Nguyen M, Trono D, Naldini L: A third-generation lentivirus vector with a conditional packaging system. J Virol 1998;72:8463–8471.

63 D'Costa J, Brown H, Kundra P, Davis-Warren A, Arya S: Human immunodeficiency virus type 2 lentiviral vectors: packaging signal and splice donor in expression and encapsidation. J Gen Virol 2001;82:425–434.

64 Poeschla EM, Wong-Staal F, Looney DJ: Efficient transduction of nondividing human cells by feline immunodeficiency virus lentiviral vectors. Nat Med 1998;4:354–357.

65 Mitrophanous K, Yoon S, Rohll J, Patil D, Wilkes F, Kim V, Kingsman S, Kingsman A, Mazarakis N: Stable gene transfer to the nervous system using a non-primate lentiviral vector. Gene Ther 1999;6:1808–1818.

66 Mangeot PE, Negre D, Dubois B, Winter AJ, Leissner P, Mehtali M, Kaiserlian D, Cosset FL, Darlix JL: Development of minimal lentivirus vectors derived from simian immunodeficiency virus (SIVmac251) and their use for gene transfer into human dendritic cells. J Virol 2000;74:8307–8315.

67 Rabson AB, Martin MA: Molecular organization of the AIDS retrovirus. Cell 1985;40: 477–480.

68 Poznansky M, Lever A, Bergeron L, Haseltine W, Sodroski J: Gene transfer into human lymphocytes by a defective human immunodeficiency virus type 1 vector. J Virol 1991;65:532–536.

69 Pluta K, Diehl W, Zhang XY, Kutner R, Bialkowska A, Reiser J: Lentiviral vectors encoding tetracycline-dependent repressors and transactivators for reversible knockdown of gene expression: a comparative study. BMC Biotechnol 2007;7:41.

70 Bachstetter AD, Jernberg J, Schlunk A, Vila JL, Hudson C, Cole MJ, Shytle RD, Tan J, Sanberg PR, Sanberg CD, Borlongan C, Kaneko Y, Tajiri N, Gemma C, Bickford PC: Spirulina promotes stem cell genesis and protects against LPS induced declines in neural stem cell proliferation. PLoS One 2010;5:e10496.

71 Jahner D, Stuhlmann H, Stewart CL, Harbers K, Lohler J, Simon I, Jaenisch R: De novo methylation and expression of retroviral genomes during mouse embryogenesis. Nature 1982;298:623–628.

72 Mitchell RS, Beitzel BF, Schroder AR, Shinn P, Chen H, Berry CC, Ecker JR, Bushman FD: Retroviral DNA integration: ASLV, HIV, and MLV show distinct target site preferences. PLoS Biol 2004;2:E234.

73 Wu X, Li Y, Crise B, Burgess SM: Transcription start regions in the human genome are favored targets for MLV integration. Science 2003;300:1749–1751.

74 Brown BD, Naldini L: Exploiting and antagonizing microRNA regulation for therapeutic and experimental applications. Nat Rev Genet 2009;10:578–585.

75 Marumoto T, Tashiro A, Friedmann-Morvinski D, Scadeng M, Soda Y, Gage FH, Verma IM: Development of a novel mouse glioma model using lentiviral vectors. Nat Med 2009;15:110–116.

76 Pfeifer A, Kessler T, Yang M, Baranov E, Kootstra N, Cheresh DA, Hoffman RM, Verma IM: Transduction of liver cells by lentiviral vectors: analysis in living animals by fluorescence imaging. Mol Ther 2001;3:319–322.

77 Sander WE, Metzger ME, Morizono K, Bonifacino A, Penzak SR, Xie YM, Chen IS, Bacon J, Sestrich SG, Szajek LP, Donahue RE: Noninvasive molecular imaging to detect transgene expression of lentiviral vector in nonhuman primates. J Nucl Med 2006;47:1212–1219.

78 Lai CM, Lai YK, Rakoczy PE: Adenovirus and adeno-associated virus vectors. DNA Cell Biol 2002;21:895–913.

79 Willadsen SM: Nuclear transplantation in sheep embryos. Nature 1986;320:63–65.

80 Rauschhuber C, Noske N, Ehrhardt A: New insights into stability of recombinant adenovirus vector genomes in mammalian cells. Eur J Cell Biol 2012;91:2–9.

81 Benihoud K, Yeh P, Perricaudet M: Adenovirus vectors for gene delivery. Curr Opin Biotechnol 1999;10:440–447.

82 Maione D, Della Rocca C, Giannetti P, D'Arrigo R, Liberatoscioli L, Franlin LL, Sandig V, Ciliberto G, La Monica N, Savino R: An improved helper-dependent adenoviral vector allows persistent gene expression after intramuscular delivery and overcomes preexisting immunity to adenovirus. Proc Natl Acad Sci USA 2001;98:5986–5991.

83 Waddington SN, McVey JH, Bhella D, Parker AL, Barker K, Atoda H, Pink R, Buckley SM, Greig JA, Denby L, Custers J, Morita T, Francischetti IM, Monteiro RQ, Barouch DH, van Rooijen N, Napoli C, Havenga MJ, Nicklin SA, Baker AH: Adenovirus serotype 5 hexon mediates liver gene transfer. Cell 2008;132:397–409.

84 Kim IH, Jozkowicz A, Piedra PA, Oka K, Chan L: Lifetime correction of genetic deficiency in mice with a single injection of helper-dependent adenoviral vector. Proc Natl Acad Sci USA 2001;98:13282–13287.

85 Toietta G, Mane VP, Norona WS, Finegold MJ, Ng P, McDonagh AF, Beaudet AL, Lee B: Lifelong elimination of hyperbilirubinemia in the Gunn rat with a single injection of helper-dependent adenoviral vector. Proc Natl Acad Sci USA 2005;102:3930–3935.

86 Brunetti-Pierri N, Stapleton GE, Law M, Breinholt J, Palmer DJ, Zuo Y, Grove NC, Finegold MJ, Rice K, Beaudet AL, Mullins CE, Ng P: Efficient, long-term hepatic gene transfer using clinically relevant HDAd doses by balloon occlusion catheter delivery in nonhuman primates. Mol Ther 2009;17:327–333.

87 Hoggan MD: Adenovirus associated viruses. Prog Med Virol 1970;12:211–239.

88 Berns KI: Parvovirus replication. Microbiol Rev 1990;54:316–329.

89 Berns KI, Giraud C: Biology of adeno-associated virus. Curr Top Microbiol Immunol 1996;218:1–23.

90 Russell DW, Hirata RK: Human gene targeting by viral vectors. Nat Genet 1998;18:325–330.

91 Miller DG, Wang PR, Petek LM, Hirata RK, Sands MS, Russell DW: Gene targeting in vivo by adeno-associated virus vectors. Nat Biotechnol 2006;24:1022–1026.

92 Hirata RK, Russell DW: Design and packaging of adeno-associated virus gene targeting vectors. J Virol 2000;74:4612–4620.

93 Hirata R, Chamberlain J, Dong R, Russell DW: Targeted transgene insertion into human chromosomes by adeno-associated virus vectors. Nat Biotechnol 2002;20:735–738.

94 Rogers CS, Stoltz DA, Meyerholz DK, Ostedgaard LS, Rokhlina T, Taft PJ, Rogan MP, Pezzulo AA, Karp PH, Itani OA, Kabel AC, Wohlford-Lenane CL, Davis GJ, Hanfland RA, Smith TL, Samuel M, Wax D, Murphy CN, Rieke A, Whitworth K, Uc A, Starner TD, Brogden KA, Shilyansky J, McCray PB Jr, Zabner J, Prather RS, Welsh MJ: Disruption of the CFTR gene produces a model of cystic fibrosis in newborn pigs. Science 2008;321:1837–1841.

95 Clark KJ, Carlson DF, Foster LK, Kong BW, Foster DN, Fahrenkrug SC: Enzymatic engineering of the porcine genome with transposons and recombinases. BMC Biotechnol 2007;7:42.

96 Doyle A, McGarry MP, Lee NA, Lee JJ: The construction of transgenic and gene knockout/knockin mouse models of human disease. Transgenic Res 2012;21:327–349.

97 Lombardo A, Genovese P, Beausejour CM, Colleoni S, Lee YL, Kim KA, Ando D, Urnov FD, Galli C, Gregory PD, Holmes MC, Naldini L: Gene editing in human stem cells using zinc finger nucleases and integrase-defective lentiviral vector delivery. Nat Biotechnol 2007;25:1298–1306.

98 Porteus MH: Mammalian gene targeting with designed zinc finger nucleases. Mol Ther 2006;13:438–446.

99 Santoni de Sio FR, Gritti A, Cascio P, Neri M, Sampaolesi M, Galli C, Luban J, Naldini L: Lentiviral vector gene transfer is limited by the proteasome at postentry steps in various types of stem cells. Stem Cells 2008;26:2142–2152.

100 Liu Y, Valadon P, Schnitzer JE: Construction of metabolically biotinylated adenovirus with deleted fiber knob as targeting vector. Virol J 2010;7:316.

101 Mowa MB, Crowther C, Arbuthnot P: Therapeutic potential of adenoviral vectors for delivery of expressed RNAi activators. Expert Opin Drug Deliv 2010;7:1373–1385.

102 Maier P, Spier I, Laufs S, Veldwijk MR, Fruehauf S, Wenz F, Zeller WJ: Chemoprotection of human hematopoietic stem cells by simultaneous lentiviral overexpression of multidrug resistance 1 and O-methylguanine-DNA methyltransferase(P140K). Gene Ther 2010; 17:389–399.

103 Ayuso E, Mingozzi F, Montane J, Leon X, Anguela XM, Haurigot V, Edmonson SA, Africa L, Zhou S, High KA, Bosch F, Wright JF: High AAV vector purity results in serotype- and tissue-independent enhancement of transduction efficiency. Gene Ther 2010;17:503–510.

104 Redaelli M, Mucignat-Caretta C, Cavaggioni A, Caretta A, D'Avella D, Denaro L, Cavirani S, Donofrio G: Bovine herpesvirus 4 based vector as a potential oncolytic-virus for treatment of glioma. Virol J 2010;7:298.

105 Warden C, Tang Q, Zhu H: Herpesvirus BACs: past, present, and future. J Biomed Biotechnol 2011;2011:124595.

106 Zhang Y, Liu Y, Zhu G, Qiu Y, Peng B, Yin J, Liu W, He X: Foamy virus: an available vector for gene transfer in neural cells and other nondividing cells. J Neurovirol 2010;16:419–426.

107 Kim DS, Nam JH: Application of attenuated coxsackievirus B3 as a viral vector system for vaccines and gene therapy. Hum Vaccin 2011;7:410–416.

108 Pichon C, Billiet L, Midoux P: Chemical vectors for gene delivery: uptake and intracellular trafficking. Curr Opin Biotechnol 2010;21:640–645.

109 Tros de Ilarduya C, Sun Y, Duzgunes N: Gene delivery by lipoplexes and polyplexes. Eur J Pharm Sci 2010;40:159–170.

110 Elouahabi A, Ruysschaert JM: Formation and intracellular trafficking of lipoplexes and polyplexes. Mol Ther 2005;11:336–347.

111 Boussif O, Lezoualc'h F, Zanta MA, Mergny MD, Scherman D, Demeneix B, Behr JP: A versatile vector for gene and oligonucleotide transfer into cells in culture and in vivo: polyethylenimine. Proc Natl Acad Sci USA 1995;92:7297–7301.

112 Pires P, Simoes S, Nir S, Gaspar R, Duzgunes N, Pedroso de Lima MC: Interaction of cationic liposomes and their DNA complexes with monocytic leukemia cells. Biochim Biophys Acta 1999;1418:71–84.

113 Simoes S, Slepushkin V, Pires P, Gaspar R, de Lima MP, Duzgunes N: Mechanisms of gene transfer mediated by lipoplexes associated with targeting ligands or pH-sensitive peptides. Gene Ther 1999;6:1798–1807.

114 Cui Z, Mumper RJ: Chitosan-based nanoparticles for topical genetic immunization. J Control Release 2001;75:409–419.

115 Bhattacharya S, Bajaj A: Advances in gene delivery through molecular design of cationic lipids. Chem Commun (Camb) 2009;31:4632–4656.

116 Dang JM, Leong KW: Natural polymers for gene delivery and tissue engineering. Adv Drug Deliv Rev 2006;58:487–499.

117 Leong KW, Mao HQ, Truong-Le VL, Roy K, Walsh SM, August JT: DNA-polycation nanospheres as non-viral gene delivery vehicles. J Control Release 1998;53:183–193.

118 Shikata F, Tokumitsu H, Ichikawa H, Fukumori Y: In vitro cellular accumulation of gadolinium incorporated into chitosan nanoparticles designed for neutron-capture therapy of cancer. Eur J Pharm Biopharm 2002;53:57–63.

119 Thanou M, Florea BI, Geldof M, Junginger HE, Borchard G: Quaternized chitosan oligomers as novel gene delivery vectors in epithelial cell lines. Biomaterials 2002;23:153–159.

120 Corsi K, Chellat F, Yahia L, Fernandes JC: Mesenchymal stem cells, MG63 and HEK293 transfection using chitosan-DNA nanoparticles. Biomaterials 2003;24:1255–1264.

121 Dastan T, Turan K: In vitro characterization and delivery of chitosan-DNA microparticles into mammalian cells. J Pharm Pharm Sci 2004;7:205–214.

122 Waterston RH, Lindblad-Toh K, Birney E, et al: Initial sequencing and comparative analysis of the mouse genome. Nature 2002;420:520–562.

123 Zhao X, Yu SB, Wu FL, Mao ZB, Yu CL: Transfection of primary chondrocytes using chitosan-pEGFP nanoparticles. J Control Release 2006;112:223–228.

124 Bishayee A, Haznagy-Radnai E, Mbimba T, Sipos P, Morazzoni P, Darvesh AS, Bhatia D, Hohmann J: Anthocyanin-rich black currant extract suppresses the growth of human hepatocellular carcinoma cells. Nat Prod Commun 2010;5:1613–1618.

125 Saranya N, Moorthi A, Saravanan S, Devi MP, Selvamurugan N: Chitosan and its derivatives for gene delivery. Int J Biol Macromol 2011;48:234–238.

126 Roy K, Mao HQ, Huang SK, Leong KW: Oral gene delivery with chitosan–DNA nanoparticles generates immunologic protection in a murine model of peanut allergy. Nat Med 1999;5:387–391.

127 Kumar M, Kong X, Behera AK, Hellermann GR, Lockey RF, Mohapatra SS: Chitosan IFN-gamma-pDNA Nanoparticle (CIN) Therapy for Allergic Asthma. Genet Vaccines Ther 2003;1:3.

128 Bowman K, Sarkar R, Raut S, Leong KW: Gene transfer to hemophilia A mice via oral delivery of FVIII-chitosan nanoparticles. J Control Release 2008;132:252–259.

129 Li GP, Liu ZG, Liao B, Zhong NS: Induction of Th1-type immune response by chitosan nanoparticles containing plasmid DNA encoding house dust mite allergen Der p 2 for oral vaccination in mice. Cell Mol Immunol 2009;6:45–50.

130 Wang J, Lee IL, Lim WS, Chia SM, Yu H, Leong KW, Mao HQ: Evaluation of collagen and methylated collagen as gene carriers. Int J Pharm 2004;279:115–126.

131 Truong-Le VL, Walsh SM, Schweibert E, Mao HQ, Guggino WB, August JT, Leong KW: Gene transfer by DNA-gelatin nanospheres. Arch Biochem Biophys 1999; 361:47–56.

132 Bourquin C, Wurzenberger C, Heidegger S, Fuchs S, Anz D, Weigel S, Sandholzer N, Winter G, Coester C, Endres S: Delivery of immunostimulatory RNA oligonucleotides by gelatin nanoparticles triggers an efficient antitumoral response. J Immunother 2010;33:935–944.

133 Wu GY, Wu CH: Receptor-mediated in vitro gene transformation by a soluble DNA carrier system. J Biol Chem 1987;262:4429–4432.

134 Pathak A, Patnaik S, Gupta KC: Recent trends in non-viral vector-mediated gene delivery. Biotechnol J 2009;4:1559–1572.

135 Schaffert D, Wagner E: Gene therapy progress and prospects: synthetic polymer-based systems. Gene Ther 2008;15:1131–1138.

136 Halama A, Kulinski M, Librowski T, Lochynski S: Polymer-based non-viral gene delivery as a concept for the treatment of cancer. Pharmacol Rep 2009;61:993–999.

137 Chollet P, Favrot MC, Hurbin A, Coll JL: Side-effects of a systemic injection of linear polyethylenimine-DNA complexes. J Gene Med 2002;4:84–91.

138 Juliano RL, Ming X, Nakagawa O, Xu R, Yoo H: Integrin targeted delivery of gene therapeutics. Theranostics 2011;1:211–219.

139 Felgner PL, Gadek TR, Holm M, Roman R, Chan HW, Wenz M, Northrop JP, Ringold GM, Danielsen M: Lipofection: a highly efficient, lipid-mediated DNA-transfection procedure. Proc Natl Acad Sci USA 1987;84: 7413–7417.

140 Huang QD, Zhong GX, Zhang Y, Ren J, Fu Y, Zhang J, Zhu W, Yu XQ: Cyclen-Based Cationic Lipids for Highly Efficient Gene Delivery towards Tumor Cells. PLoS One 2011;6:e23134.

141 Yellayi S, Hilliard B, Ghazanfar M, Tsingalia A, Nantz MH, Bollinger L, de Kok-Mercado F, Hecker JG: A single intrathecal injection of DNA and an asymmetric cationic lipid as lipoplexes ameliorates experimental autoimmune encephalomyelitis. Mol Pharm 2011;8:1980–1984.

142 Whittaker RH, Margulis L: Protist classification and the kingdoms of organisms. Biosystems 1978;10:3–18.

143 Steidler L: Gene exchange of thyA for interleukin-10 secures live GMO bacterial therapeutics. Discov Med 2003;3:49–51.

144 Steidler L, Hans W, Schotte L, Neirynck S, Obermeier F, Falk W, Fiers W, Remaut E: Treatment of murine colitis by *Lactococcus lactis* secreting interleukin-10. Science 2000; 289:1352–1355.

145 Steidler L, Rottiers P: Therapeutic drug delivery by genetically modified Lactococcus lactis. Ann NY Acad Sci 2006;1072:176–186.

146 Bermudes D, Zheng LM, King IC: Live bacteria as anticancer agents and tumor-selective protein delivery vectors. Curr Opin Drug Discov Dev 2002;5:194–199.

147 Cunningham C, Nemunaitis J: A phase I trial of genetically modified Salmonella typhimurium expressing cytosine deaminase (TAPET-CD, VNP20029) administered by intratumoral injection in combination with 5-fluorocytosine for patients with advanced or metastatic cancer. Protocol No. CL-017. Version: April 9, 2001. Hum Gene Ther 2001; 12:1594–1596.

148 Brenner S: The genetics of *Caenorhabditis elegans*. Genetics 1974;77:71–94.

149 Silverman GA, Luke CJ, Bhatia SR, Long OS, Vetica AC, Perlmutter DH, Pak SC: Modeling molecular and cellular aspects of human disease using the nematode *Caenorhabditis elegans*. Pediatr Res 2009;65:10–18.

150 Sulston JE, Horvitz HR: Post-embryonic cell lineages of the nematode, *Caenorhabditis elegans*. Dev Biol 1977;56:110–156.

151 Ellis HM, Horvitz HR: Genetic control of programmed cell death in the nematode C. elegans. Cell 1986;44:817–829.

152 Golstein P, Kroemer G: Cell death by necrosis: towards a molecular definition. Trends Biochem Sci 2007;32:37–43.

153 Link CD: Expression of human beta-amyloid peptide in transgenic *Caenorhabditis elegans*. Proc Natl Acad Sci USA 1995;92:9368–9372.

154 Gidalevitz T, Ben-Zvi A, Ho KH, Brignull HR, Morimoto RI: Progressive disruption of cellular protein folding in models of polyglutamine diseases. Science 2006;311:1471–1474.

155 Kwok TC, Ricker N, Fraser R, Chan AW, Burns A, Stanley EF, McCourt P, Cutler SR, Roy PJ: A small-molecule screen in *C. elegans* yields a new calcium channel antagonist. Nature 2006;441:91–95.

156 Banno Y, Shimada T, Kajiura Z, Sezutsu H: The silkworm-an attractive BioResource supplied by Japan. Exp Anim 2010;59:139–146.

157 Daubnerova I, Roller L, Zitnan D: Transgenesis approaches for functional analysis of peptidergic cells in the silkworm *Bombyx mori*. Gen Comp Endocrinol 2009;162:36–42.

158 Alphey L, Nimmo D, O'Connell S, Alphey N: Insect population suppression using engineered insects. Adv Exp Med Biol 2008;627:93–103.

159 Marshall JM, Taylor CE: Malaria control with transgenic mosquitoes. PLoS Med 2009;6:e20.

160 Chen CH, Huang H, Ward CM, Su JT, Schaeffer LV, Guo M, Hay BA: A synthetic maternal-effect selfish genetic element drives population replacement in *Drosophila*. Science 2007;316:597–600.

161 Thomas DD, Donnelly CA, Wood RJ, Alphey LS: Insect population control using a dominant, repressible, lethal genetic system. Science 2000;287:2474–2476.

162 Haffter P, Nusslein-Volhard C: Large scale genetics in a small vertebrate, the zebrafish. Int J Dev Biol 1996;40:221–227.

163 Kawakami K: Transposon tools and methods in zebrafish. Dev Dyn 2005;234:244–254.

164 Feitsma H, Cuppen E: Zebrafish as a cancer model. Mol Cancer Res 2008;6:685–694.

165 Milan DJ, Macrae CA: Zebrafish genetic models for arrhythmia. Prog Biophys Mol Biol 2008;98:301–308.

166 Buckley CE, Goldsmith P, Franklin RJ: Zebrafish myelination: a transparent model for remyelination? Dis Model Mech 2008;1:221–228.

167 Chesneau A, Sachs LM, Chai N, Chen Y, Du Pasquier L, Loeber J, Pollet N, Reilly M, Weeks DL, Bronchain OJ: Transgenesis procedures in *Xenopus*. Biol Cell 2008;100:503–521.

168 Etkin LD, Pearman B: Distribution, expression and germ line transmission of exogenous DNA sequences following microinjection into *Xenopus laevis* eggs. Development 1987;99:15–23.

169 Amaya E, Kroll KL: A method for generating transgenic frog embryos. Methods Mol Biol 1999;97:393–414.

170 Smith K, Spadafora C: Sperm-mediated gene transfer: applications and implications. Bioessays 2005;27:551–562.

171 Sinzelle L, Vallin J, Coen L, Chesneau A, Du Pasquier D, Pollet N, Demeneix B, Mazabraud A: Generation of trangenic *Xenopus laevis* using the *Sleeping Beauty* transposon system. Transgenic Res 2006;15: 751–760.

172 Pan FC, Chen Y, Loeber J, Henningfeld K, Pieler T: I-SceI meganuclease-mediated transgenesis in *Xenopus*. Dev Dyn 2006;235: 247–252.

173 Allen BG, Weeks DL: Using phiC31 integrase to make transgenic *Xenopus laevis* embryos. Nat Protoc 2006;1:1248–1257.

174 Richardson P, Chapman J: The *Xenopus tropicalis* genome project. • Current Genomics 2003;4:645–652.

175 Wells DE, Gutierrez L, Xu Z, Krylov V, Macha J, Blankenburg KP, Hitchens M, Bellot LJ, Spivey M, Stemple DL, Kowis A, Ye Y, Pasternak S, Owen J, Tran T, Slavikova R, Tumova L, Tlapakova T, Seifertova E, Scherer SE, Sater AK: A genetic map of *Xenopus tropicalis*. Dev Biol 2011;354:1–8.

176 Stubbs JL, Davidson L, Keller R, Kintner C: Radial intercalation of ciliated cells during *Xenopus* skin development. Development 2006;133:2507–2515.

177 Geng X, Xiao L, Lin GF, Hu R, Wang JH, Rupp RA, Ding X: Lef/Tcf-dependent Wnt/ beta-catenin signaling during *Xenopus* axis specification. FEBS Lett 2003;547:1–6.

178 Moritz OL, Biddle KE, Tam BM: Selection of transgenic *Xenopus laevis* using antibiotic resistance. Transgenic Res 2002;11:315–319.

179 Fowler CD, Arends MA, Kenny PJ: Subtypes of nicotinic acetylcholine receptors in nicotine reward, dependence, withdrawal: evidence from genetically modified mice. Behav Pharmacol 2008;19:461–484.

180 Sanchez A, Ramirez P, Pino G, Chavez R, Majado M, Munitiz V, Munoz A, Palenciano CG, Yelamos J, Rodriguez-Gago M, Pons JA, Parrilla P: Immunopathology of an hDAF transgenic pig model liver xenotransplant into a primate. Transplant Proc 2003;35:2041–2042.

181 Adams DH, Kadner A, Chen RH, Farivar RS: Human membrane cofactor protein (MCP, CD 46) protects transgenic pig hearts from hyperacute rejection in primates. Xenotransplantation 2001;8:36–40.

182 Diamond LE, Quinn CM, Martin MJ, Lawson J, Platt JL, Logan JS: A human CD46 transgenic pig model system for the study of discordant xenotransplantation. Transplantation 2001;71:132–142.

183 Loveland BE, Milland J, Kyriakou P, Thorley BR, Christiansen D, Lanteri MB, Regensburg M, Duffield M, French AJ, Williams L, Baker L, Brandon MR, Xing PX, Kahn D, McKenzie IF: Characterization of a CD46 transgenic pig and protection of transgenic kidneys against hyperacute rejection in non-immunosuppressed baboons. Xenotransplantation 2004;11:171–183.

184 Schmoeckel M, Bhatti FN, Zaidi A, Cozzi E, Waterworth PD, Tolan MJ, Pino-Chavez G, Goddard M, Warner RG, Langford GA, Dunning JJ, Wallwork J, White DJ: Orthotopic heart transplantation in a transgenic pig-to-primate model. Transplantation 1998;65:1570–1577.

185 Waterworth PD, Dunning J, Tolan M, Cozzi E, Langford G, Chavez G, White D, Wallwork J: Life-supporting pig-to-baboon heart xenotransplantation. J Heart Lung Transplant 1998;17:1201–1207.

186 Kuang H, Wang PL, Tsien JZ: Towards transgenic primates: what can we learn from mouse genetics? Sci China C Life Sci 2009;52:506–514.

187 Yang SH, Cheng PH, Banta H, Piotrowska-Nitsche K, Yang JJ, Cheng EC, Snyder B, Larkin K, Liu J, Orkin J, Fang ZH, Smith Y, Bachevalier J, Zola SM, Li SH, Li XJ, Chan AW: Towards a transgenic model of Huntington's disease in a non-human primate. Nature 2008;453:921–924.

188 Sasaki E, Suemizu H, Shimada A, Hanazawa K, Oiwa R, Kamioka M, Tomioka I, Sotomaru Y, Hirakawa R, Eto T, Shiozawa S, Maeda T, Ito M, Ito R, Kito C, Yagihashi C, Kawai K, Miyoshi H, Tanioka Y, Tamaoki N, Habu S, Okano H, Nomura T: Generation of transgenic non-human primates with germline transmission. Nature 2009;459:523–527.

189 Schatten G, Mitalipov S: Developmental biology: transgenic primate offspring. Nature 2009;459:515–516.

190 Chan AW, Yang SH: Generation of transgenic monkeys with human inherited genetic disease. Methods 2009;49:78–84.

Pascale Piguet
Pharmazentrum, University of Basel
Klingelbergstrasse 50
CH–4056 Basel (Switzerland)
Tel. +41 61 267 16 14, E-Mail pascale.piguet@unibas.ch

Piguet P, Poindron P (eds): Genetically Modified Organisms and Genetic Engineering in Research and Therapy. BioValley Monogr. Basel, Karger, 2012, vol 3, pp 33–38

Restriction Enzymes: From Their Discovery to Their Applications

Werner Arber

Biozentrum, University of Basel, Basel, Switzerland

Abstract

As a contribution to the history of science, the discovery and the functions of bacterial restriction endonucleases, as well as their applications in molecular genetic analysis and in biotechnological innovations, including genetic engineering, are described here from the personal viewpoint of the author.   Copyright © 2012 S. Karger AG, Basel

It was already known in the 1920s that particular viruses, called bacteriophages or shortly phages, propagate in bacteria. Upon successful infection, the phage propagates and after an hour or so the bacteria break open by lysis and liberate the phage progeny consisting of one to two hundreds of phage particles per bacterial host cell. Based on this knowledge, the hope to use phage as an antibiotic against pathogenic bacteria in a series of trials never yielded reliable results.

Each specific phage strain has its particular host range, i.e. the phage can successfully infect a limited number of different bacterial host strains. This often depends on genetically controlled incompatibilities of various kinds. Host range mutations can affect either the phage or a potential bacterial host. In the 1950s a few independent reports described still another phenomenon affecting the host range of phages, called host-controlled modification. As long as a phage is propagated in the same host strain, practically each infection is successful. But when an alternative potential host becomes infected the success rate can, for example, be only one of 10,000 infected cells. Interestingly, the rare phage liberated by the alternative host now infects successfully this latest host strain, but not any longer with full efficiency its previous host. Upon such back and forth shifting of host bacteria, the ability of successful infection of the parental host gets each time lost and the one to infect the new host gets acquired. This is unlikely to be due to host range mutations. As was already mentioned, this phenomenon became first known as host-controlled modification and later as restriction and modification.

The molecular mechanisms of restriction and modification became unraveled in the 1960s. At that time, we undertook at the University of Geneva investigations on radiation effects on microorganisms, bacteria and phages. In the course of this work with several strains of bacteria, we encountered the restriction and modification phenomenon in our experiments. We could thereby observe that upon restriction, the radioactively labeled DNA of the restricted infecting phage became rapidly acid soluble, pointing to its degradation. We thus raised the question if modification also affects the phage DNA in contrast to an alternative hypothesis that modified phage particles carried a host gene product that would enable the phage to infect the same host again. At that time, other researchers had just studied the semiconservative replication of DNA molecules. Luckily, two hosts were available for our research: one with a restriction modification system and the other without. This allowed us to grow modified phage just for one growth cycle in the nonrestricting, nonmodifying host strain. By different labeling strategies of the parental phage DNA, we could easily follow the infectivity of progeny phage with either semiconserved or after high multiplicity of infection also fully conserved parental DNA. These experiments clearly showed that parental modification allowed phage with semiconserved as well as conserved DNA to reinfect their previous host, while all phage particles with newly synthesized double-stranded DNA had become restricted for their previous host. A few years later, it became clear that modification enabling a phage to successfully infect a given host is an enzymatically mediated methylation of a nucleotide carried in a relatively short specific nucleotide sequence, which we call the recognition site. Clearly, the methyl group carried on a nucleotide in each DNA strand of the recognition site does not affect biological functions encoded in the concerned DNA segments. From today's point of view, host-controlled modification was one of the very first molecularly unraveled epigenetic phenomena.

In the mid-1960s, available knowledge on molecular mechanisms of restriction and modification suggested that these bacterial systems serve to limit the rates of acquisition of foreign genetic information to low levels, whereas they allow a more efficient exchange of genetic information to occur between closely related bacterial strains carrying the same restriction and modification system. This idea found good experimental support: indeed, restriction and modification affects not only phage DNA upon infection, but also bacterial DNA upon conjugation and upon transformation.

When, in the late 1960s, the first restriction endonucleases and, soon thereafter, modification methylases were purified and used in enzymatic studies in vitro, their expected functions as endonucleases and as DNA methylases, respectively, became confirmed. Work carried out in the 1970s in several laboratories (including the Biozentrum of the University of Basel) revealed that independent restriction and modification systems could have different molecular compositions and could also use different mechanistic approaches to reach the same results. These consist, on the one hand, of the cleavage of invading foreign DNA into fragments that afterwards become further degraded by independent exonucleases and, on the other hand, in protecting the cell's own DNA from restriction cleavage by applying site-specific methylation. For example, type I restriction enzymes cleave foreign DNA after specific recognition rather randomly outside of the short recognition sequences. In contrast, type II restriction enzymes cleave foreign DNA reproducibly at the recognition sites. Around 1970, it became obvious that type II enzymes could well serve for the structural and functional analysis of genomes.

It is good to remind in this context that the genome is defined as the entire set of genetic information of a living organism. This includes, besides the classical open reading frames serving for protein synthesis, DNA segments of various lengths which may exert still largely unknown biological functions. In a paradigmatic comparison between the linear sequences of nucleotide building blocks of DNA and the linear sequences of letters in our written language, an open reading frame of average length is composed of about 1,000 letters which corresponds to a relatively short paragraph. The bacterial genome corresponds to one book, depending on the kind of bacteria of a small or a large book. For example, the genome of *E. coli* bacteria has roughly the size of the Bible. Higher eukaryotic organisms, animals and plants carry genomes corresponding to an encyclopedia of several hundred up to 1,000 volumes.

For structural and functional studies, one has to sort out relatively short genomic segments and amplify them in order to have enough material available for analysis. This can be done by genetic engineering, in which a selected DNA fragment is spliced into a vector DNA molecule (often a viral genome or a plasmid). After introduction of the resulting auto-replicating hybrid into an appropriate host cell, the DNA segment under study can become replicated and carried functions may, at least sometimes, also get expressed.

At a scientific workshop on restriction and modification systems that we organized in 1972 at the Meeting Center Leuenberg near Basel, some participants proposed to also discuss risks that could be inherent to the prospective field of genetic engineering, including horizontal transfer of genomic information into other kinds of organisms. This debate had its first follow-up in a letter of a group of American scientists to Science, requesting a conference devoted to this kind of conjectural risks. The suggested international conference was held in February 1975 in Asilomar, Calif., USA. The Conference participants agreed to distinguish between short-term potential risks such as toxicity, allergenic reactions or other noxious effects, on the one hand, and on the other hand, longer term, evolutionary risks that could occur after accidental or deliberate release of recombinant DNA into the environment, if the involved DNA would, at some later time, become horizontally transferred to other organisms. The Conference concluded that guidelines for work with recombinant DNA should be introduced in order to avoid (or at least to minimize) any undesirable effects of this kind of experimentation.

As a participant at the Asilomar Conference, I became aware that a more profound scientific knowledge on molecular mechanisms of spontaneously occurring genetic variation, the driving force of biological evolution, was needed in order to compare natural genetic variation with engineered genetic variation.

In the meantime, we have obtained good insight into natural genetic variation, in particular from experimental work with bacteria and bacteriophages. More recently, major results found good support for their general validity by computational DNA sequence comparisons between more or less closely related genomes including those of animals and plants. In the following section, we will shortly present the current scientific knowledge on spontaneously occurring genetic variation.

Genetic variation in nature is an active process in which specific gene products – acting as variation generators and/or as modulators of the rates of genetic variation – are involved. They work hand in hand with nongenetic elements such as intrinsic properties of matter and of molecules (e.g. isomeric conformations), effects of chemical and physical mutagens and random encounter. We have become aware that a multitude of specific molecular mechanisms contribute to the overall genetic variation. These specific mechanisms can be assigned to three natural strategies of genetic variation: (1) Small local changes in the genomic DNA sequences. These include the substitution of a nucleotide by another one, the additional insertion of one or a few adjacent nucleotides, the deletion of one or a few adjacent nucleotides and the scrambling of a few adjacent nucleotides. (2) A second natural strategy of genetic variation brings about a structural rearrangement of usually relatively short segments of the genomic DNA. These mostly enzyme-mediated reshuffling processes can result in the duplication, deletion, inversion or translocation of a genomic segment. (3) The third natural strategy of spontaneous genetic variation consists of the acquisition of a segment of foreign DNA by horizontal transfer. In the bacterial world, an efficient limitation of this process to low, tolerable rates is brought about by the restriction enzymes. Acquisition of a short cleavage segment of foreign DNA is occasionally ensured by a fast integration into the recipient genome. Not the least, thanks to the universality of the genetic code, DNA acquisition is a highly effective evolutionary step that allows the recipient organism to profit from a long evolutionary development done elsewhere. Local sequence changes and intragenomic DNA rearrangements contribute with different qualities to functional developments in the course of evolutionary progress. For example, nucleotide substitution may by chance bring about an improvement of an already existing function. DNA rearrangements can, also by chance, cause the fusion of two previously independent functional domains which can result in a novel kind of function. Or the fusion of an open reading frame with an alternative control signal for gene expression can result in an altered availability of the concerned gene product.

We have to be aware that any kind of genetic variant will be submitted to natural selection, which is based on the ability of an organism to deal with its encountered environment, i.e. with the material composition of its habitat and with the activities of the other living organisms populating the same habitat.

These considerations led us to compare the natural evolution with designed steps of evolution by genetic engineering. First of all, any functional alteration brought about by genetic engineering will also become submitted to the effects of natural selection. Conversely, any changes that genetic engineering may cause to the genome of an organism are based on one of the three strategies described above to work in natural, spontaneous genetic variation. For example, site-directed mutagenesis can bring about a local sequence change; intragenomic deletion, translocation and amplification of a DNA segment by genetic engineering are comparable with naturally occurring DNA rearrangements, and the splicing of a segment of foreign DNA into a genome is comparable with the naturally occurring DNA acquisition by horizontal transfer. The only difference between natural events and genetic engineering is that the latter alterations are usually planned by the investigator. From long-term observations, we know that in the natural evolutionary progress and in classical breeding experiments, undesirable effects are quite rare. Since genetic engineering uses the same three strategies of genetic variation as those acting in spontaneous genetic variation, conjectural risks of genetic engineering must be comparable to those of natural genetic variation, i.e. quite low. Remember that any functional DNA segment found in nature and transferred horizontally to another organism by genetic engineering must have had good chances in the past times to become transferred by natural means to other types of organisms, and this might also occur in future times.

In view of these considerations, we can conclude that genetic engineering does respect the laws of nature that ensure the natural progress of evolution. There is no scientific reason to assign great, particular risks to the method of genetic engineering. On the basis of ethical considerations, we as humans can ameliorate our bases and conditions of living as long as we fully respect the laws of nature and the established ethical norms. For example, the production of genetically engineered food plants (GM crops) entirely obeys these rules. This also holds for microorganisms engineered to produce particular organic compounds or in view of their use for bioremediation. For ethical reasons, microorganisms should not be genetically engineered to render them stronger pathogens. Firm ethical norms should be followed for any deliberate genetic alteration in animals and in particular in humans.

The wide use of bacterial restriction endonucleases of many different recognition site specificities has brought about in the past few decades important impulses to molecular genetic investigations and to the analysis, as well as to the applications of biological functions. The thereby acquired scientific knowledge on genomics, proteomics and molecular evolution also has its philosophical values and deserves to belong to the cultural heritage of mankind.

Selected Reading

Arber W: Host specificity of DNA produced by *Escherichia coli*. V. The role of methionine in the production of host specificity. J Mol Biol 1965;11:247–256.

Arber W: Host-controlled modification of bacteriophage. Annu Rev Microbiol 1965;19:365–378.

Arber W: DNA modification and restriction. Prog Nucleic Acid Res Mol Biol 1974;14:1–37.

Arber W: Promotion and limitation of genetic exchange. Science 1979;205:361–365.

Arber W: Elements for a theory of molecular evolution. Gene 2003;317:3–11.

Arber W: Genetic variation and molecular evolution; in Meyers RA (ed.): Genomics and Genetics. Weinheim, Wiley-VCH, 2007, vol 1, pp 385–406.

Arber W, Dussoix D: Host specificity of DNA produced by *Escherichia coli*. I. Host-controlled modification of bacteriophage lambda. J Mol Biol 1962;5:18–36.

Arber W, Linn S: DNA modification and restriction. Annu Rev Biochem 1969;38:467–500.

Arber W, Morse ML: Host specificity of DNA produced by *Escherichia coli*. VI. Effects on bacterial conjugation. Genetics 1965;51:137–148.

Bickle TA, Krüger DH: Biology of DNA restriction. Microbiol Rev 1993;57:434–450.

Dussoix D, Arber W: Host specificity of DNA produced by *Escherichia coli*. II. Control over acceptance of DNA from infecting phage lambda. J Mol Biol 1962;5:37–49.

Roberts RJ, Vincze T, Posfai J, Macelis D: REBASE – a database for DNA restriction and modification: enzymes, genes and genomes. Nucleic Acids Res 2010;38:D234–D236.

Smith HO, Wilson RW: A restriction enzyme from *Hemophilus influenzae*. I. Purification and general properties. J Mol Biol 1970;51:379–391.

Prof. Werner Arber
Biozentrum, University of Basel
Klingelgerbstrasse 70
CH–4056 Basel (Switzerland)
Tel. +41 61 267 21 30, E-Mail Werner.Arber@unibas.ch

Piguet P, Poindron P (eds): Genetically Modified Organisms and Genetic Engineering in Research and Therapy. BioValley Monogr. Basel, Karger, 2012, vol 3, pp 39–49

Transforming Growth Factor-Beta Superfamily: Animal Models for Development and Disease

Björn Spittau[a] · Eleni Roussa[a] · Klaus Unsicker[a] · Kerstin Krieglstein[a,b]

[a]Institute of Anatomy and Cell Biology and [b]FRIAS-Lifenet, University of Freiburg, Freiburg, Germany

Abstract
Transforming growth factor-beta (TGF-β) comprises multifunctional extracellular signaling molecules acting in the nervous system from early development into adulthood. The TGF-β superfamily constitutes a group of 33 members. Many of these execute important functions during early embryogenesis, organogenesis, after birth and in the adult, as well as in tissue repair and homeostasis. Manipulating their expression in several species has generated valuable animal models to dissect their precise function. This overview will concentrate on the role of TGF-β isoforms TGF-β1, TGF-β2 and TGF-β3, and on GDF-15 in the nervous system.

Transforming growth factor-beta (TGF-β) regulates numerous cell functions in the developing and adult brain. TGF-βs are secreted dimeric proteins that signal via heteromeric transmembrane serine-threonine kinase receptors. Phosphorylation of receptor-Smads leads to the formation of complexes with the common-Smad4, which translocates to the nucleus to regulate in a larger transcriptional complex immediate early gene and target gene expression [1, 2]. Three different isoforms, TGF-β1, TGF-β2, and TGF-β3, are expressed in mammalian tissues; the isoforms TGF-β2 and TGF-β3 are expressed in the developing nervous system [3, 4]. The TGF-β superfamily of proteins includes 33 members in mammals [5]. They are divided into two major subgroups, the first one comprising, inter alia, TGF-βs, activins, nodal and myostatin/GDF8, Mullerian-inhibiting substance, and a second group including the bone morphogenetic proteins (BMP) and growth/differentiation factors (GDF). TGF-β/BMP-like proteins are found in vertebrates and invertebrates, including *Caenorhabditis elegans* and *Drosophila melanogaster* [6]. Many of these signaling proteins execute important functions during early embryogenesis, organogenesis, after birth and in the adult, as well as in tissue repair and homeostasis [7–9].

Mouse models, based on target ablations of several components of the TGF-β signaling pathway, have been instrumental in revealing important functions in development and adult maintenance (see for example [10] and the most recent special issue of *Cell & Tissue Research* on TGF-β in aging and disease [11]).

In the nervous system, TGF-βs exert roles in neurons and glia and are involved in the regulation of proliferation, migration, differentiation, survival and death [12–14].

TGF-β in Neuronal Differentiation and Synaptogenesis

TGF-β has been implicated in the regulation of neurite outgrowth, transmitter synthesis as well as synapse formation. TGF-β has been reported to cause neurite sprouting and elongation of hippocampal axons as well as promoting re-elongation of injured axons of hippocampal neurons in vitro [15, 16] and to initiate axon formation and neuronal migration in the developing neocortex in vivo [17]. Extracellular signaling factors such as Wnt and TGF-βs are recognized as target-derived signals in synaptogenesis of the invertebrate neuromuscular junction [18, 19]. In the past years all components of the TGF-β signaling system have been localized in the presynaptic terminal of the neuromuscular junction, whereby TGF-β ligands are synthesized and localized on the postsynaptic side. In chick ciliary ganglionic neurons developmental expression of K_{Ca} channels coincides with synaptogenesis. Dryer et al. [20] have shown that target-derived TGF-β1 regulates the developmental expression of Ca^{2+}-activated K^+ channels in vitro and in vivo. The acute effect of TGF-β1 relies on the translocation of K_{Ca} channels from intracellular stores to the plasma membrane involving signaling via RAS, Erk, and PI3 kinase [20]. TGF-β is also known to have a prominent role in long-term synaptic facilitation in isolated Aplysia ganglia [21]. Within minutes, TGF-β1 stimulated MAPK-dependent phosphorylation of synapsin, which appeared to modulate synapsin distribution, and resulted in a reduced magnitude of synaptic depression [22]. Most recently, Fukushima et al. [23] were able to show that TGF-β modulates synaptic efficacy and plasticity in dissociated rat hippocampal neurons. Together, increasing evidence suggests that TGF-β may be involved in synaptogenesis, modulation of synaptic transmission and synaptic plasticity.

To study TGF-β-dependent synaptogenesis and neuronal network development, mice deficient in TGF-β2 were analyzed [24]. TGF-β2-deficient mice die within one hour after birth due to congenital cyanosis. Cardiovascular and pulmonary causes of lethality had already been excluded by others [25]. To test whether functional failures in the brainstem respiratory network might cause the TGF-β2-KO phenotype, we analyzed excitatory and inhibitory synaptic transmission of neurons in the pre-Bötzinger complex (preBötC), which is an essential part of the central respiratory rhythm-generating network.

We investigated, in collaboration with Zhang and Sargsyan [24], the overall network activity of neurons in the pre-BötC. Acute slices containing the pre-BötC and the hypoglossal nucleus were used for whole cell recordings. Both frequency and amplitude of spontaneous PSCs of preBötC neurons were impaired in TGF-β2-KO mice. Recordings of pharmacologically isolated spontaneous GABAergic/glycinergic and glutamatergic postsynaptic currents revealed impaired synaptic transmission of both systems. This may be explained either by a reduced number of synapses or by compromised neurotransmitter release machinery in all synapses in the region of the preBötC [24]. The latter hypothesis has been strengthened by demonstrating that hypertonic stimulation also resulted in a reduced total charge transfer, suggesting that the deletion of TGF-β2 mainly impaired the presynaptic component of both the inhibitory and excitatory synaptic transmission in all synapses in the region of the preBötC (for further details, see [24]).

To address the question of impaired synaptogenesis as a cause of reduced synaptic transmission, synapses were identified by immunohistochemistry using synaptophysin and synapsin as general synaptic markers, vesicular glutamate transporter (vGlut2) as a marker for excitatory glutamatergic presynaptic terminals, and vesicular GABA transporter (vGat) as a marker for inhibitory GABAergic presynaptic terminals. These experiments were conducted in collaboration with Varoqueaux [24]. Numbers of synaptophysin-positive presynaptic terminals in the preBötC in TGF-β2 null mice at E18.5 were significantly increased to about double the numbers as compared to wild-type littermates. Numbers of synapsin and vGlut2-positive synapses were also increased, while numbers of vGat positive synapses were unchanged. To test for impaired synapse formation, the expression levels of a selection of the most relevant synaptic proteins were analyzed in TGF-β2 KO mice and wild-type littermates in the brainstem and in entire brain homogenates by means of quantitative Western blot analysis [cf. 24]). One of the few proteins with a change in expression was vGLUT1, which was increased in brainstem material in E18.5 TGF-β2 KO mice as compared to wild-type littermates. These data are in line with the increased number of perisomatic synapses in TGF-β2 mutant mice.

Function of TGF-β2 in Hippocampal Neuronal Network Maturation

To test whether TGF-β2-dependent neuronal network maturation is a more general phenomenon, we extended our analysis to other neuronal networks as well, primarily to the hippocampus. The respiratory system in the brainstem is already developed and functional at birth, while other networks including the hippocampus will only mature within the first postnatal weeks. As TGF-β2 KO mice die at birth, we established and used cultures of mouse dissociated hippocampi at E18.5 and allowed them to mature in vitro for 2 to 3 weeks [26, 27]. As shown by Lacmann et al. [26], hippocampal neurons mature and form a neuronal network in vitro. Furthermore, neuronal activity regulates TGF-β release and expression and upregulates TGF-β-inducible early gene (Tieg1) expression, demonstrating that activity-dependent released TGF-β may exert autocrine actions and thereby activate the TGF-β-dependent signaling pathway. Together, these results suggest an activity-dependent release and gene transcription of TGF-β from mouse hippocampal neurons in vitro as well as subsequent autocrine functions of the released TGF-β within the hippocampal network [26]. Nonetheless, TGF-β treatment does result in the regulation of a substantial number of target genes upon short term as well as long term exposure [27].

Regulation of Survival and Death

TGF-β has been shown to promote neuron survival of several neuron populations in vitro [28]. For eliciting its neurotrophic actions TGF-β cooperates and modulates the neurotrophic capacities of numerous growth factors including neurotrophins [29] and, most importantly, GDNF [30]. GDNF was shown to crucially depend on TGF-β to exert its neurotrophic activities on peripheral as well as mesencephalic dopaminergic neurons [31]. In vivo, its neuroprotective effect on target-deprived preganglionic sympathetic neurons also depends on the presence of TGF-β [32]. GDNF/TGF-β cooperativity on chick ciliary ganglionic neurons has now been characterized in detail [33, 34], evidence was provided that TGF-β is required for appropriate GDNF receptor recruitment to the plasma membrane.

Depending on the cellular context, TGF-β has also been shown to regulate ontogenetic neuron death. Upon immunoneutralization of all TGF-β isoforms in ovo (E6–E10) ontogenetic cell death of chick parasympathetic ciliary ganglionic neurons, sensory dorsal root ganglionic neurons as well as lumbar spinal motoneurons could be prevented [35]. Similarly, TGF-β regulates ontogenetic morphogenetic cell death in the developing retina of chick and mouse embryos [36, 37]. Furthermore, induced neuron death following embryonic limb bud ablation in chick embryos resulted in a significant neuroprotection upon immunoneutralization of TGF-β [35]. Another classical model for morphogenetic cell death during embryogenesis represents the removal of interdigital tissue to form individual fingers. Similarly, double deletion of TGF-β2 and -β3 in the mouse resulted in lack of cell death [38]. Further along this line, apoptosis was significantly reduced in the intestinal mucosa of Tgf-β2(+/−) and Tgf-β3(+/−) heterozygous mice. This decrease in apoptosis was accompanied by an increase

                                    Spittau · Roussa · Unsicker · Krieglstein

in villus length and regulation of apoptosis-associated proteins Bcl-xL and Bcl-2; proliferation, however, seemed to remain unchanged [39]. Together, these data suggest TGF-β as a key regulator of ontogenetic cell death in vivo.

Although TGF-β induced apoptosis and underlying signaling pathways have been well characterized in many cells types, little is known on the signaling pathways triggered in TGF-β-induced apoptosis in the nervous system [40–43]. TGF-β-induced cell death was accompanied by a loss of Bcl-xL protein [44]. Loss of Bcl-xL may be regulated via repression of its transcription through the TGF-β immediate early genes Tieg1 or Tieg3, also referred to as Klf10 and Klf11 [41, 45, 46]. Furthermore, by performing GST-pull down experiments using Bcl-xl as a bait we were able to identify Fractin as a Bcl-xL interacting protein in the course of TGF-β-induced cell death [43].

Neuroimmunology

TGF-βs play essential roles in different tissues and cell types under physiological and pathological conditions. Especially TGF-β1 has been shown to be a major immunoregulatory molecule with further direct neuroprotective properties. TGF-β1 is expressed in the healthy central nervous system and its upregulation has been reported for a variety of neurodegenerative disease models. Genetic targeting of TGF-βs in mice [10] revealed that only TGF-β1 mutant mice show a phenotype that links TGF-β1 to neuroimmunology. Mice with homozygous deletion for TGF-β1 develop normally without any gross abnormalities. However, three weeks after birth, these animals die due to a systemic inflammatory response which leads to tissue necrosis, organ failure and subsequent death [47]. The inflammatory response in TGF-β1 mutant mice further affects the central nervous system. Brionne et al. [48] reported that lack of TGF-β1 in neonatal mice resulted in a dramatic increase of neurodegeneration which was accompanied by a reduction in synaptophysin and laminin expression. Moreover, TGF-β1 mutant mice show prominent microglia activation – referred to as microgliosis – that could either be causative for the increased neurodegeneration in TGF-β1 mutants, or the consequence of neuronal death in these animals. A straightforward attempt has been made by Makwana et al. [49]. In their study, the authors used TGF-β1 mutants which were bred on a T and B cell-deficient RAG2$^{-/-}$ background to avoid the lethal systemic inflammatory response. These TGF-β1 mutants survived into adulthood and revealed an extensive inflammatory response in the brain. Increased astroglial GFAP and CD44 expression as well as pronounced microglial proliferation, high expression of phagocytotic and inflammatory markers and reduced branching of microglial extensions were observed.

TGF-βs are secreted as homodimers which are bound to latency associated proteins (LAPs) that keep TGF-βs in an inactive state. These complexes further interact with extracellular matrix components, such as latent TGF-β-binding proteins (LTBPs) or integrins which are important for the extracellular storage of TGF-βs and the regulation of TGF-β activation by thrombospondin-1, plasmin, MMP-2 and MMP-9 [50]. Mice with mutant TGF-β1, unable to interact with LTBP1 [51] and thrombospondin-1-deficient mice [52] both develop tumors and inflammatory responses in several organs, underlining the importance of TGF-β activation in the extracellular space. Although the availability of bioactive TGF-β1 is important for the regulation of neuroinflammatory responses, excessive TGF-β signaling in the central nervous system seems to have detrimental effects. Transgenic mice overexpressing bioactive TGF-β1 under the astroglia-specific GFAP promoter develop a communicating hydrocephalus due to increased production of extracellular matrix components such as laminin and fibronectin [53]. Moreover, TGF-β1-overexpressing mice are more susceptible to the rodent MS model experimental autoimmune encephalomyelitis, showing an early-onset, increased disease progression and enhanced infiltration of mononuclear cells [54].

Lessons from mutant mice clearly taught us that TGF-β1 is an important modulator of neuroimmunologic responses and its expression, extracellular storage and activation have to be tightly regulated processes. However, TGF-β1 and its receptors are expressed by virtually all cell types of the central nervous system under physiological and pathophysiological conditions, thus posing an exciting challenge to gain insight into how TGF-β1 regulates developmental processes and contributes to the course of neurodegenerative pathologies such as Parkinson's disease.

Growth Differentiation Factor-15

Growth differentiation factor-15 (GDF-15) is a relatively novel member of the TGF-β superfamily that was independently discovered by several groups following different strategies. We identified GDF-15 by screening EST databases for conserved consensus sequences of TGF-βs [55, 56; see also for gene and protein structure]. Research on GDF-15 in the past decade has mainly focused on its diverse roles in human diseases and cancer biology [cf. 57, 58]. However, relatively little is known on the physiological roles of GDF-15, even in tissue and organs such as liver, lung and kidney, where it is highly expressed. Interesting functions of GDF-15 outside the cancer field include its implications in myocardial infarction and its property as a biomarker for risk stratification of mortality in chronic heart failure, after acute coronary syndrome and other cardiac diseases [59–61]. Furthermore, its protective role against atherosclerosis is of utmost clinical relevance: in atherosclerosis, GDF-15 inhibits CCR2-mediated chemotaxis of macrophages and necrotic core formation, a process which is also TβRII-dependent [62].

                                                    Spittau · Roussa · Unsicker · Krieglstein

Focusing on the putative roles of GDF-15 in the nervous system, we have shown that GDF-15 is widely distributed in the CNS and PNS [63]. Using qRT-PCR and Western blotting, low levels (when compared to liver, lung, kidney) of GDF-15 mRNA and protein are found in all regions of the unlesioned rat and mouse CNS, in peripheral nerves, in isolated astrocytes and dorsal root ganglion cells (DRGs). Highest levels of GDF-15 are expressed in the choroid plexus; the protein is secreted into the CSF. GDF-15 turned out to be a very potent neurotrophic factor for dopaminergic (DAergic) neurons cultured from the embryonic rat midbrain and for 6-OHDA-lesioned DAergic neurons in the adult substantia nigra, with an efficacy exceeding that of GDNF.

To begin to understand implications of GDF-15 in the orchestration of brain lesions [64], we investigated the regulation of GDF-15 subsequent to a cold lesion of the cortex. We found GDF-15 mRNA to be highly upregulated in regions adjacent to the lesion site, with a peak at 36–48 h. In lesion conditions, GDF-15 is prominently expressed by neurons and less by microglial cells. Similar patterns of GDF-15 upregulation were detected in the hippocampus and cortex following a transient occlusion of the middle cerebral artery [65].

The GDF-15 knockout has revealed an interesting phenotype further corroborating the notion that GDF-15 may serve as a major neurotrophic factor for several populations of neurons in the CNS and PNS [66]. A progressive numerical decline was noted in spinal and brainstem motoneurons, which began to develop during the fourth postnatal month reaching its peak at the age of 6 months, with a loss of 20% of the neuron populations. A similar decline occurred in DRG neurons, while sympathetic neurons were not affected. Neuron losses were accompanied by losses of motor axons and impairment of rotarod skills. Schwann cells are a major source of GDF-15 in peripheral nerves; the cells secrete GDF-15, and axons transport it retrogradely to the cell soma. Despite striking similarities in the phenotypes of the GDF-15 and CNTF knockouts, expression levels of CNTF and other neurotrophic factors in the sciatic nerve were unaltered.

The receptor(s) for GDF-15 have not been unequivocally identified. Several studies have suggested that GDF-15 might signal through canonical TGF-β receptors and signaling (i.e. Smad) pathways. Using cardiomyocytes transformed with SMAD-decoy oligonucleotides, Heger et al. [67] observed a blockade of GDF-15 mediated protein synthesis and hypertrophy. In addition, also inhibitors of PI3-kinase and ERK blocked the hypertrophic growth response to GDF-15. In accordance with this study, the Breit group [68] reported that the effects of MIC-1/GDF-15 on hypothalamic arcuate neurons, which mediate tumor-induced anorexia and weight loss, could be blocked with antibodies to TβRII. However, studies in our laboratory (Peterziel, Paech, personal communication) so far failed to confirm GDF-15 signaling through TGF-β (or GDNF) receptors. Even so, we also identified phosphorylated Akt and ERK as downstream components of GDF-15 mediated signaling on cerebellar granule neurons [69].

Acknowledgements

Work of our laboratories has been funded through the Deutsche Forschungsgemeinschaft, for example through SFBs 406, 488, 503, 592, 780, and individual research grants to K.K. and K.U., through the BMBF, EraNet Neuron, and through the Excellence Initiative (FRIAS).

References

1   Moustakas A, Heldin CH: The regulation of TGFbeta signal transduction. Development 2009;136:3699–3714.

2   Shi Y, Massagué J: Mechanisms of TGF-beta signaling from cell membrane to the nucleus. Cell 2003;113:685–700.

3   Roberts AB, Sporn MB: The transforming growth factor-βs; in Sporn MB, Roberts AB (eds): Handbook of Experimental Pharmacology. Heidelberg, Springer Verlag, 1990, vol 95, pp 419–472.

4   Flanders KC, Ludecke G, Engels S, Cissel DS, Roberts AB, Kondaiah P, Lafyatis R, Sporn MB, Unsicker K: Localization and actions of transforming growth factor-betas in the embryonic nervous system. Development 1991;113:183–191.

5   Derynck R, Miyazono K: TGF-β and the TGF-β family; in Derynck R, Miyazono K (eds): The TGF-β Family. Cold Spring Harbor, Cold Spring Harbor Laboratory Press, 2008, pp 29–43.

6   Kahlem P, Newfeld SJ: Informatics approaches to understanding TGFbeta pathway regulation. Development 2009;136:3729–3740.

7   Hogan BL: Bone morphogenetic proteins: multifunctional regulators of vertebrate development. Genes Dev 1996;10:1580–1594.

8   Chang H, Brown CW, Matzuk MM: Genetic analysis of the mammalian transforming growth factor-beta superfamily. Endocr Rev 2002;23:787–823.

9   Umulis D, O'Connor MB, Blair SS: The extracellular regulation of bone morphogenetic protein signaling. Development 2009;136:3715–3728.

10   Dünker N, Krieglstein K: Targeted mutations of transforming growth factor-beta genes reveal important roles in mouse development and adult homeostasis. Eur J Biochem 2000;267:6982–6988.

11   Krieglstein K, Miyazono K, Ten Dijke P, Unsicker K: TGF-β in aging and disease. Cell Tissue Res 2012;347:5–9.

12   Böttner M, Krieglstein K, Unsicker K: The transforming growth factor-betas: structure, signaling, and roles in nervous system development and functions. J Neurochem 2000;75:2227–2240.

13   Krieglstein K: Tranforming growth factor-betas in the brain. Handb Neurochem and Mol Neurobiol 2006, pp 123–141.

14   Krieglstein K, Zheng F, Unsicker K, Alzheimer C: More than being protective: functional roles for TGF-β/activin signaling pathways at central synapses. Trends Neurosci 2011;34:421–429.

15   Ishihara A, Saito H, Abe K: Transforming growth factor-beta 1 and -beta 2 promote neurite sprouting and elongation of cultured rat hippocampal neurons. Brain Res 1994;39:21–25.

16   Abe K, Chu PJ, Ishihara A, Saito H: Transforming growth factor-beta 1 promotes re-elongation of injured axons of cultured rat hippocampal neurons. Brain Res 1996;723:206–209.

17   Yi JJ, Barnes AP, Hand R, Polleux F, Ehlers MD: TGF-beta signaling specifies axons during brain development. Cell 2010;142:144–157.

18   Salinas PC: Signaling at the vertebrate synapse: new roles for embryonic morphogens? J Neurobiol 2005;64:435–445.

   Spittau · Roussa · Unsicker · Krieglstein

19 Packard M, Mathew D, Budnik V: Wnts and
TGF beta in synaptogenesis: old friends
signalling at new places. Nat Rev Neurosci
2003;4:113–120.
20 Dryer SE, Lhuillier L, Cameron JS, Martin-
Caraballo M: Expression of K(Ca) channels
in identified populations of developing vertebrate neurons: role of neurotrophic factors
and activity. J Physiol Paris 2003;97:49–58.
21 Zhang F, Endo S, Cleary LJ, Eskin A, Byrne
JH: Role of transforming growth factor-beta
in long-term synaptic facilitation in Aplysia.
Science 1997;275:1318–1320.
22 Chin J, Angers A, Cleary LJ, Eskin A, Byrne
JH: Transforming growth factor beta1
alters synapsin distribution and modulates
synaptic depression in Aplysia. J Neurosci
2002;22:RC220.
23 Fukushima T, Liu RY, Byrne JH:
Transforming growth factor-beta2 modulates
synaptic efficacy and plasticity and induces
phosphorylation of CREB in hippocampal
neurons. Hippocampus 2007;17:5–9.
24 Heupel K, Sargsyan V, Plomp JJ, Rickmann
M, Varoqueaux F, Zhang W, Krieglstein K:
Loss of transforming growth factor-beta 2
leads to impairment of central synapse function. Neural Dev 2008;3:25.
25 Sanford LP, Ormsby I, Gittenberger-de
Groot AC, Sariola H, Friedman R, Boivin
GP, Cardell EL, Doetschman T: TGFbeta2
knockout mice have multiple developmental defects that are non-overlapping
with other TGFbeta knockout phenotypes.
Development 1997;124:2659–2670.
26 Lacmann A, Hess D, Gohla G, Roussa E,
Krieglstein K: Activity-dependent release
of transforming growth factor-beta in a
neuronal network in vitro. Neuroscience
2007;150:647–657.
27 Vogel T, Ahrens S, Büttner N, Krieglstein K:
Transforming growth factor beta promotes
neuronal cell fate of mouse cortical and hippocampal progenitors in vitro and in vivo:
identification of Nedd9 as an essential signaling component. Cereb Cortex 2010;20:661–
671.
28 Krieglstein K, Strelau J, Schober A, Sullivan
A, Unsicker K: TGF-beta and the regulation
of neuron survival and death. J Physiol Paris
2002;96:25–30.
29 Krieglstein K, Unsicker K: Distinct modu-
latory actions of TGF-beta and LIF on
neurotrophin-mediated survival of developing sensory neurons. Neurochem Res
1996;21:843–850.
30 Krieglstein K, Henheik P, Farkas L, Jaszai
J, Galter D, Krohn K, Unsicker K: Glial cell
line-derived neurotrophic factor requires
transforming growth factor-beta for exerting
its full neurotrophic potential on peripheral
and CNS neurons. J Neurosci 1998;18:9822–
9834.
31 Schober A, Peterziel H, von Bartheld CS,
Simon H, Krieglstein K, Unsicker K: GDNF
applied to the MPTP-lesioned nigrostriatal
system requires TGF-beta for its neuroprotective action. Neurobiol Dis 2007;25:378–
391.
32 Schober A, Hertel R, Arumäe U, Farkas L,
Jaszai J, Krieglstein K, Saarma M, Unsicker
K: Glial cell line-derived neurotrophic factor rescues target-deprived sympathetic
spinal cord neurons but requires transforming growth factor-beta as cofactor in vivo.
J Neurosci 1999;19:2008–2015.
33 Peterziel H, Unsicker K, Krieglstein K:
Molecular mechanisms underlying the
cooperative effect of glial cell line-derived
neurotrophic factor and transforming
growth factor beta in neurons. J Cell Biol
2002;159:157–169.
34 Peterziel H, Paech T, Strelau J, Unsicker K,
Krieglstein K: Specificity in the crosstalk of
TGFbeta/GDNF family members is determined by distinct GFR alpha receptors.
J Neurochem 2007;103:2491–2504.
35 Krieglstein K, Richter S, Farkas L, Schuster
N, Dünker N, Oppenheim RW, Unsicker
K: Reduction of endogenous transforming
growth factors beta prevents ontogenetic
neuron death. Nat Neurosci 2000;3:1085–
1090.
36 Dünker N, Schuster N, Krieglstein K:
Transforming growth factor beta modulates
programmed cell death in the retina of the
developing chick embryo. Development
2001;128:1933–1942.
37 Dünker N, Krieglstein K: Genetic evidence
for transforming growth factor beta-
mediated programmed cell death in the
developing mouse retina. Cell Tissue Res
2003;313:1–10.

Piguet P, Poindron P (eds): Genetically Modified Organisms and Genetic Engineering in Research and Therapy. BioValley Monogr. Basel, Karger, 2012, vol 3, pp 50–59

Transfection of Human Neuroblastoma Cells with Alzheimer's Disease Brain Hallmarks as a Promising Strategy to Investigate the Role of Neurosteroidogenesis in Neuroprotection

Christine Patte-Mensah[a] · Laurence Meyer[a] · Véronique Schaeffer[a] · Anne Eckert[b] · Ayikoe G. Mensah-Nyagan[a]

[a]Equipe Stéroïdes, Neuromodulateurs et Neuropathologies, EA-4438, Université de Strasbourg, Bâtiment 3 de la Faculté de Médecine, Strasbourg, France; [b]Neurobiology Laboratory for Brain Aging and Mental Health, Psychiatric University Clinic, Basel, Switzerland

Abstract

The human neuroblastoma SH-SY5Y cell line is known to be a relevant cellular model for biochemical investigations on Alzheimer's disease (AD). SH-SY5Y cells were also characterized as neurosteroid-producing cells expressing key steroidogenic enzymes. Therefore, SH-SY5Y cells were used to develop an original strategy aimed at determining whether neurosteroidogenesis may be an endogenous mechanism involved in the protection against neurodegenerative processes. The first step was to stably transfect SH-SY5Y cells with key proteins of AD such as the human native tau (hTau40), mutant tau (P301L) and the wild-type amyloid precursor protein (APPwt). The second step consisted of combining pulse-chase experiments, high-performance liquid chromatography (HPLC) and continuous flow scintillation to investigate neurosteroid synthesis in SH-SY5Y native cells compared to cells stably transfected with hTau40, P301L, APPwt or control vectors. While microscopic analyses revealed that AD protein overexpression did not affect the morphology of SH-SY5Y cells, biochemical assays showed that hTau40 increased neurosteroid synthesis from the precursor pregnenolone. The pathogenic P301L mutation suppressed the stimulatory action exerted by Tau on neurosteroidogenesis and APPwt overexpression induced a selective effect depending on each step of neurosteroidogenic pathways. The data show that genetic engineering in SH-SY5Y cells allowed determination of a significant impact exerted on neurosteroid synthesis by AD-associated peptides. AD pathogenic factors may therefore induce neurodegeneration by decreasing in nerve cells endogenous production of neuroprotective neurosteroids. The technological approach used herein appears as a promising strategy to get valuable insights into the clarification of pathophysiological mechanisms involved in neurodegenerative diseases.

The high levels of proteins involved in the formation of plaques and neuro-fibrillary tangles in Alzheimer's disease (AD) were correlated with decreased cerebral concentrations of neurosteroids such as pregnenolone sulfate (PREGS) and dehydroepiandrosterone sulfate or DHEAS [1, 2]. Neurosteroid biosynthesis (neurosteroidogenesis) has been evidenced in the nervous system of several animal species and in humans indicating that this process may control important neurophysiological functions [3–8]. Several investigations using rodents as models revealed that the synthesis within the nervous system of progesterone (PROG) and its active metabolites, dihydroprogesterone (5α-DHP) and tetrahydroprogesterone (3α,5α-THP), contributes to the protection of nerve cells against degeneration [8–11]. However, the role of neurosteroidogenesis in the regulation of degenerative mechanisms in humans is not clarified yet and this situation hampers the development of effective neurosteroid-based neuroprotective therapies. In AD and various other human neurodegenerative diseases, the main pathologic components of neurofibrillary tangles are paired helical filaments of highly phosphory-lated microtubule-associated protein tau [12–15]. Unlike normal tau which maintains a microtubule structure, hyperphosphorylated tau sequesters pro-teins associated to microtubules and provokes destabilization of the neuronal cytoskeleton leading to neurodegeneration [16]. The common tau missense mutation associated with neurodegenerative disorders in humans is Pro[301]-to-Leu (P301L tau) which leads to intracellular neurofibrillary lesions composed of hyperphosphorylated tau in frontotemporal dementia with Parkinsonism linked to chromosome 17 [17–19]. AD is also characterized by an amyloido-genic pathway leading to the formation of β-amyloid plaques which constitute with neurofibrillary tangles the pivotal lesions. Although a progress has been made over the past years regarding the involvement of the amyloid precursor protein (APP) in AD, the biological activity of this protein remains a mat-ter of speculation [20–23]. It was demonstrated that APP binds to the brain-specific signal-transducing Go protein and APP mutants of familial AD can induce Go-mediated apoptosis in neurons [24–25].

In order to determine whether key proteins involved in the modulation of neurodegenerative processes may interfere with neurosteroid biosynthesis in humans, an original protein transfection-based approach was designed to inves-tigate the effects of the wild-type APP (APPwt), the human native tau (hTau40) and mutant P301L tau on neurosteroidogenesis in SH-SY5Y cells. This cell line, which is considered the most representative cellular model for investigations on AD [26–34], was also characterized as a neurosteroid-producing cell line containing various key steroid-synthesizing enzymes [35–37]. Therefore, the well-validated approach combining pulse-chase experiments, high-performance liquid chromatography (HPLC) and continuous flow scintillation detection [38–45] was used to investigate neurosteroid synthesis in SH-SY5Y native cells compared to cells stably transfected with hTau40, P301L, APPwt or control vectors.

Cell Culture, Transfection and Microscope Analyses

Human neuroblastoma SH-SY5Y cells were grown at 37°C under an atmosphere of 5% CO_2 in DMEM supplemented with 10% (v/v) heat-inactivated fetal calf serum, 5% (v/v) heat-inactivated horse serum, 2 mM Glutamax and 1% (v/v) penicillin/streptomycin. SH-SY5Y cells were stably transfected with DNA constructs harboring human wild-type APP_{695} (APPwt) or the expression vector pCEP4 (Invitrogen, Europe) alone (control vector) using lipofect AMINEplus [46]. Transfected APPwt cells were grown in DMEM standard medium supplemented with 300 µg/ml hygromycin. SH-SY5Y cells were also stably transfected with either the longest 4-repeat isoform of hTau40 or mutant P301L human tau cDNA constructs or empty vector (pRc/RSV, Invitrogen, Europe) using lipofect AMINE2000 [33]. These cells were grown in DMEM standard medium supplemented with 300 µg/ml G418. In all experiments, stably transfected SH-SY5Y cells were compared with cells overexpressing the corresponding empty or control vectors. Cells were passaged every 3–4 days and were used for pulse-chase experiments when they reached 80–90% confluence.

Thanks to western blot analyses, the stably transfected SH-SY5Y cells (compared to native neuroblastoma cells or to control cells transfected with empty vectors) have been well characterized as overexpressing the corresponding protein, i.e. the wild-type hTau40, the pathogenic P301L tau or APPwt [33, 46].

Morphological Analysis

For the morphological analysis, cells were seeded at a density of 2.5×10^5 cells/cm^2 on coverslips previously coated with 0.05 mg/ml collagen and examined under a DMR microscope equipped with a digital camera assisted by a Pentium IV PC computer (Leica, Wetzlar, Germany).

The native human neuroblastoma SH-SY5Y cells exhibited a neuroblast-like morphology with well-differentiated perikarya (arrows) and occasional short neurites (arrowheads) as shown in figure 1. Stable transfection of SH-SY5Y cells with the expression vector pRc/RSV, which was either empty or harboring expression constructs encoding the wild-type hTau40 or the pathogenic P301L

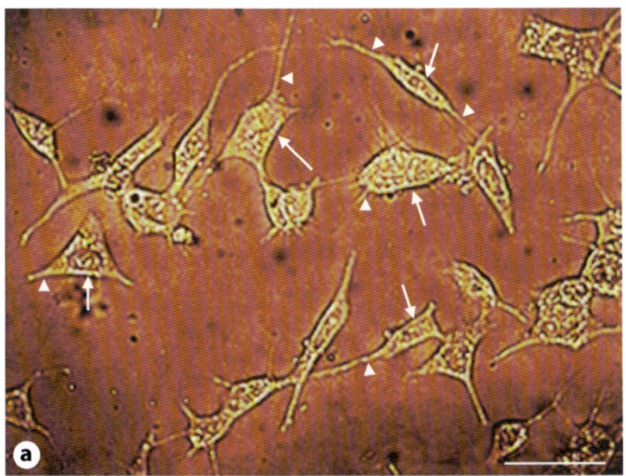

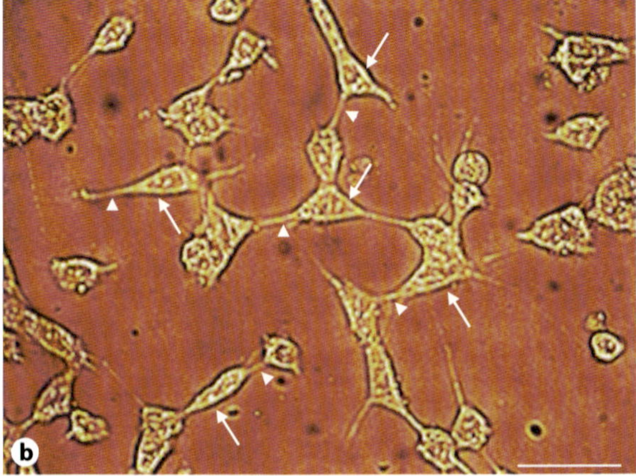

Fig. 1. Morphological analysis of native and transfected SH-SY5Y cells. **a** Native SH-SY5Y cells exhibited a neuroblast-like morphology with differentiated perikarya (arrows) and occasional short neurites (arrowheads). **b** Transfection of SH-SY5Y cells with the empty vector or with the vector containing cDNA encoding hTau40, mutant P301L tau and APPwt did not significantly change the morphological aspect of the cells. Scale bars = 50 μm.

mutant tau did not modify the morphological aspect of the cells. Similarly, there was no detectable changes in the morphology of SH-SY5Y cells transfected with the pCEP4 vector alone or harboring the APPwt cDNA (fig. 1a, b).

Neurosteroidogenesis in SH-SY5Y Cells

The biosynthesis of neurosteroids was investigated in native and stably transfected SH-SY5Y cells by determining their capacity to convert pregnenolone (the precursor of all classes of steroids) into steroidal metabolites. The conventional

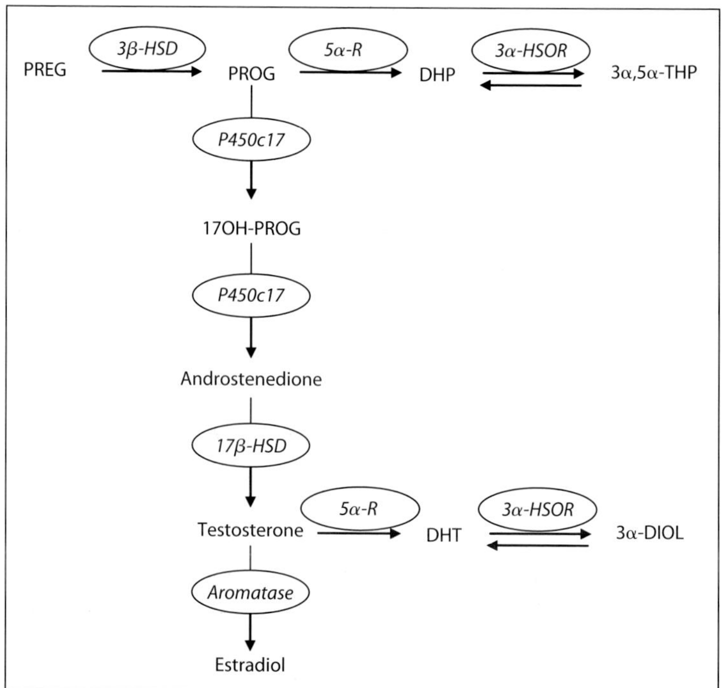

Fig. 2. Biochemical pathways of neurosteroids from the precursor pregnenolone. 3β-HSD = 3β-Hydroxysteroid dehydrogenase; P450c17 = cytochrome P450c17 or 17α-hydroxylase/17,20 lyase; 17β-HSD = 17β-hydroxysteroid dehydrogenase; 5α-R = 5α-reductase; 3α-HSOR = 3α-hydroxysteroid oxidoreductase; PREG = pregnenolone; PROG = progesterone; 17OH-PROG = 17α-hydroxyprogesterone; DHP = dihydroprogesterone; 3α,5α-THP, 3α,5α-tetrahydroprogesterone; DHT = dihydrotestosterone; 3α-DIOL = 3α-androstanediol.

steroid biosynthetic pathway from the precursor pregnenolone is recapitulated in figure 2.

A 3-hour incubation of native or stably transfected SH-SY5Y cells with [³H] PREG yielded the formation of several tritiated metabolites. Reversed-phase HPLC analysis coupled with Flo/One characterization of steroids showed that the main [³H]-steroids newly synthesized from the precursor [³H]PREG in all categories of cells were [³H]PROG, [³H]3α,5α-THP, [³H]17OH-PROG, [³H]T, [³H]3α-DIOL and [³H]estradiol (fig. 2). Quantitative analyzes revealed no difference in neurosteroid production in native and vector control cells. Overexpression of hTau40 increased neurosteroid synthesis de novo from the precursor pregnenolone. The pathogenic P301L mutation suppressed the stimulatory action exerted by Tau on neurosteroidogenesis. Transfection of APPwt induced a selective effect depending on each biosynthetic step of neurosteroidogenic pathways. Indeed, overexpression of APPwt inhibited progesterone formation, unaffected

    Patte-Mensah · Meyer · Schaeffer · Eckert · Mensah-Nyagan

17-hydroxyprogesterone and testosterone but increased 3α-androstanediol and estradiol formation in SH-SY5Y cells [47].

Discussion

Thanks to the stable transfection of AD biomarkers in neuroblastoma cells, it was possible to observe that the human native Tau (hTau40) strongly increased the ability of SH-SY5Y cells to newly synthesize PROG, 17OH-PROG, T and 3α-DIOL from the precursor PREG [33, 47]. Unlike hTau40, the mutant tau P301L is totally devoid of significant action on neurosteroid production in SH-SY5Y cells. The wild-type tau protein, which acts as a promoter of tubulin polymerization, is involved in axonal transport and, by inhibiting the rate of microtubule depolymerization, it contributes to the stabilization of the cytoskeleton and neuritic outgrowth [48–52]. The most obvious pathological event in several neurodegenerative disorders is abnormal phosphorylation of tau proteins leading to their aggregation into intraneuronal filamentous inclusions [14]. Transgenic mice overexpressing the P301L mutant human tau protein exhibited an accumulation of hyperphosphorylated tau and developed neurofibrillary tangles [53–56]. In addition, several neurodegenerative disorders in humans are associated with P301L mutation of tau protein which provokes neurofibrillary lesions in frontotemporal dementia with Parkinsonism linked to chromosome 17 [17–19]. Therefore, the fact that the strong stimulatory action exerted by hTau40 on neurosteroidogenesis in SH-SY5Y cells cannot be reproduced by the mutant P301L tau indicates that the process of neurosteroid biosynthesis may belong to cellular mechanisms only activated by wild-type tau to stabilize the cytoskeleton and to protect nerve cells against degeneration. In support of this suggestion, elegant studies have shown that endogenous synthesis of the neurosteroid PREG in nerve cells promotes microtubule assembly and the development of cytoskeleton by increasing tubulin polymerization [57, 58]. In addition, it has been observed that the stimulatory action of hTau40 was particularly high on the neosynthesis of PROG known to be a potent neuroprotective neurosteroid in the central and the peripheral nervous systems [9]. Overexpression of hTau40 in SH-SY5Y cells has also increased the formation of 3α-DIOL which exhibited similar neuroprotective properties as PROG [59, 60]. P301L mutation-evoked disappearance of tau stimulatory effect on neurosteroidogenesis perfectly matches with proteomic and functional analyses revealing abnormal activity of mitochondria (key steroid-producing organelles) in P301L tau transgenic mice [56]. The effects of APPwt overexpression on the process of neurosteroidogenesis in SH-SY5Y cells were steroid-dependent. Indeed, expression of APPwt, which inhibited the formation of PROG from PREG, unaffected the production of 17OH-PROG and T, and stimulated the neosynthesis of 3α-DIOL and estradiol [47]. This result suggests that APPwt itself and/or the increased generation of Aβ due to the overexpression of APPwt in these cells [46] exert different actions on the enzymatic machinery responsible for neurosteroid production. Therefore, the key steroidogenic enzymes responsible for the synthesis of PROG (3β-hydroxysteroid dehydrogenase, 3β-HSD), 3α-DIOL (3α-hydroxysteroid oxidoreductase, 3α-HSOR) and estradiol (aromatase) may not have a similar sensitivity to Aβ. In agreement with this suggestion, biochemical studies demonstrated that extracellular Aβ modulates in a dose-dependent manner various neurosteroid-synthesizing pathways in SH-SY5Y cells [61].

In conclusion, the present paper shows that the overexpression of AD biomarkers did not change the morphology of SH-SY5Y cells but modified their neurosteroidogenic ability.

The results suggest that AD pathogenic factors may induce neurodegeneration through the reduction of neuroprotective neurosteroid synthesis in nerve cells. The technological methods used to perform the works including stably transfection of SH-SY5Y cells, pulse-chase, HPLC and flow scintillation detection appear as valuable tools to elucidate AD pathophysiological processes.

References

1 Weill-Engerer S, David JP, Sazdovitch V, Liere P, Eychenne B, Pianos A, Schumacher M, Delacourte A, Baulieu EE, Akwa Y: Neurosteroid quantification in human brain regions: comparison between Alzheimer's and nondemented patients. J Clin Endocrinol Metab 2002;87:5138–5143.

2 Kim SB, Hill M, Kwak YT, Hampl R, Jo DH, Morfin R: Neurosteroids: cerebrospinal fluid levels for Alzheimer's disease and vascular dementia diagnostics. J Clin Endocrinol Metab 2003;88:5199–5206.

3 Baulieu EE, Robel P, Schumacher M: Contemporary Endocrinology. Totowa, Humana Press, 1999.

4 Mensah-Nyagan AG, Do-Rego JL, Beaujean D, Luu-The V, Pelletier G, Vaudry H: Neurosteroids: expression of steroidogenic enzymes and regulation of steroid biosynthesis in the central nervous system. Pharmacol Rev 1999;51:63–81.

5 Mensah-Nyagan AG, Beaujean D, Luu-The V, Pelletier G, Vaudry H: Anatomical and biochemical evidence for the synthesis of unconjugated and sulfated neurosteroids in amphibians. Brain Res Brain Res Rev 2001; 37:13–24.

6 Mensah-Nyagan AG, Do-Rego JL, Beaujean D, Luu-The V, Pelletier G, Vaudry H: Regulation of neurosteroid biosynthesis in the frog diencephalon by GABA and endozepines. Horm Behav 2001;40:218–225.

7 Stoffel-Wagner B: Neurosteroid metabolism in the human brain. Eur J Endocrinol 2001;145:669–679.

8 Schaeffer V, Meyer L, Patte-Mensah C, Mensah-Nyagan AG: Progress in dorsal root ganglion neurosteroidogenic activity: basic evidence and pathophysiological correlation. Prog Neurobiol 2010;92:33–41.

9 Melcangi RC, Garcia-Segura LM, Mensah-Nyagan AG: Neuroactive steroids: state of the art and new perspectives. Cell Mol Life Sci 2008;65:777–797.

10 Meyer L, Patte-Mensah C, Taleb O, Mensah-Nyagan AG: Cellular and functional evidence for a protective action of neurosteroids against vincristine chemotherapy-induced painful neuropathy. Cell Mol Life Sci 2010; 67:3017–3034.

11 Meyer L, Patte-Mensah C, Taleb O, Mensah-Nyagan AG: Allopregnanolone prevents and suppresses oxaliplatin-evoked painful neuropathy: multi-parametric assessment and direct evidence. Pain 2011;152:170–181.

12 Grundke-Iqbal I, Iqbal K, Tung YC, Quinlan M, Wisniewski HM, Binder LI: Abnormal phosphorylation of the microtubule-associated protein tau (tau) in Alzheimer cytoskeletal pathology. Proc Natl Acad Sci USA 1986;83:4913–4917.

13 Kosik KS, Joachim CL, Selkoe DJ: Microtubule-associated protein tau (tau) is a major antigenic component of paired helical filaments in Alzheimer disease. Proc Natl Acad Sci USA 1986;83:4044–4048.

14 Buée L, Bussière T, Buée-Scherrer V, Delacourte A, Hof PR: Tau protein isoforms, phosphorylation and role in neurodegenerative disorders. Brain Res Brain Res Rev 2000;33:95–130.

15 Iqbal K, Alonso Adel C, Chen S, Chohan MO, El-Akkad E, Gong CX, Khatoon S, Li B, Liu F, Rahman A, Tanimukai H, Grundke-Iqbal I: Tau pathology in Alzheimer disease and other tauopathies. Biochim Biophys Acta 2005;1739:198–210.

16 Alonso AD, Grundke-Iqbal I, Barra HS, Iqbal K: Abnormal phosphorylation of tau and the mechanism of Alzheimer neurofibrillary degeneration: sequestration of microtubule-associated proteins 1 and 2 and the disassembly of microtubules by the abnormal tau. Proc Natl Acad Sci USA 1997;94:298–303.

17 Hutton M, Lendon CL, Rizzu P, et al: Association of missense and 5′-splice-site mutations in tau with the inherited dementia FTDP-17. Nature 1998;393:702–705.

18 Poorkaj P, Bird TD, Wijsman E, Nemens E, Garruto RM, Anderson L, Andreadis A, Wiederholt WC, Raskind M, Schellenberg GD: Tau is a candidate gene for chromosome 17 frontotemporal dementia. Ann Neurol 1998;43:815–825.

19 Spillantini MG, Murrell JR, Goedert M, Farlow MR, Klug A, Ghetti B: Mutation in the tau gene in familial multiple system tauopathy with presenile dementia. Proc Natl Acad Sci USA 1998;95:7737–7741.

20 Mattson MP: Cellular actions of beta-amyloid precursor protein and its soluble and fibrillogenic derivatives. Physiol Rev 1997;77:1081–1132.

21 Bayer TA, Cappai R, Masters CL, Beyreuther K, Multhaup G: It all sticks together – the APP-related family of proteins and Alzheimer's disease. Mol Psychiatry 1999;4:524–528.

22 Coulson EJ, Paliga K, Beyreuther K, Masters CL: What the evolution of the amyloid protein precursor supergene family tells us about its function. Neurochem Int 2000;36:175–184.

23 Heber S, Herms J, Gajic V, Hainfellner J, Aguzzi A, Rulicke T, von Kretzschmar H, von Koch C, Sisodia S, Tremml P, Lipp HP, Wolfer DP, Muller U: Mice with combined gene knock-outs reveal essential and partially redundant functions of amyloid precursor protein family members. J Neurosci 2000;20:7951–7963.

24 Nishimoto I, Okamoto T, Matsuura Y, Takahashi S, Murayama Y, Ogata E: Alzheimer amyloid protein precursor complexes with brain GTP-binding protein G(o). Nature 1993;362:75–79.

25 Yamatsuji T, Matsui T, Okamoto T, Komatsuzaki K, Takeda S, Fukumoto H, Iwatsubo T, Suzuki N, Asami-Odaka A, Ireland S, Kinane TB, Giambarella U, Nishimoto I: G protein-mediated neuronal DNA fragmentation induced by familial Alzheimer's disease-associated mutants of APP. Science 1996;272:1349–1352.

26 Tanaka T, Iqbal K, Trenkner E, Liu DJ, Grundke-Iqbal I: Abnormally phosphorylated tau in SY5Y human neuroblastoma cells. FEBS Lett 1995;360:5–9.

27 Li YP, Bushnell AF, Lee CM, Perlmutter LS, Wong SK: Beta-amyloid induces apoptosis in human-derived neurotypic SH-SY5Y cells. Brain Res 1996;738:196–204.

28 Mailliot C, Bussière T, Caillet-Boudin ML, Delacourte A, Buée L: Alzheimer-specific epitope of AT100 in transfected cell lines with tau: toward an efficient cell model of tau abnormal phosphorylation. Neurosci Lett 1998;255:13–16.

29 Zhong J, Iqbal K, Grundke-Iqbal I: Hyperphosphorylated tau in SY5Y cells: similarities and dissimilarities to abnormally hyperphosphorylated tau from Alzheimer disease brain. FEBS Lett 1999;453:224–228.

30 Misonou H, Morishima-Kawashima M, Ihara Y: Oxidative stress induces intracellular accumulation of amyloid beta-protein (Abeta) in human neuroblastoma cells. Biochemistry 2000;39:6951–6959.

31 Wang J, Tung YC, Wang Y, Li XT, Iqbal K, Grundke-Iqbal I: Hyperphosphorylation and accumulation of neurofilament proteins in Alzheimer disease brain and in okadaic acid-treated SY5Y cells. FEBS Lett 2001;507:81–87.

32 Olivieri G, Novakovic M, Savaskan E, Meier F, Baysang G, Brockhaus M, Muller-Spahn F: The effects of beta-estradiol on SHSY5Y neuroblastoma cells during heavy metal induced oxidative stress, neurotoxicity and beta-amyloid secretion. Neuroscience 2002;113:849–855.

33 Ferrari A, Hoerndli F, Baechi T, Nitsch RM, Götz J: Beta-amyloid induces paired helical filament-like tau filaments in tissue culture. J Biol Chem 2003;278:40162–40168.

34 Jämsä A, Hasslund K, Cowburn RF, Bäckström A, Vasänge M: The retinoic acid and brain-derived neurotrophic factor differentiated SH-SY5Y cell line as a model for Alzheimer's disease-like tau phosphorylation. Biochem Biophys Res Commun 2004; 319:993–1000.

35 Melcangi RC, Maggi R, Martini L: Testosterone and progesterone metabolism in the human neuroblastoma cell line SH-SY5Y. J Steroid Biochem Mol Biol 1993;46:811–818.

36 Wozniak A, Hutchison RE, Morris CM, Hutchison JB: Neuroblastoma and Alzheimer's disease brain cells contain aromatase activity. Steroids 1998;63:263–267.

37 Guarneri P, Cascio C, Piccoli T, Piccoli F, Guarneri R: Human neuroblastoma SH-SY5Y cell line: neurosteroid-producing cell line relying on cytoskeletal organization. J Neurosci Res 2000;60:656–665.

38 Mensah-Nyagan AG, Feuilloley M, Dupont E, Do-Rego JL, Leboulenger F, Pelletier G, Vaudry H: Immunocytochemical localization and biological activity of 3 beta-hydroxysteroid dehydrogenase in the central nervous system of the frog. J Neurosci 1994; 14:7306–7318.

39 Mensah-Nyagan AG, Do-Rego JL, Feuilloley M, Marcual A, Lange C, Pelletier G, Vaudry H: In vivo and in vitro evidence for the biosynthesis of testosterone in the telencephalon of the female frog. J Neurochem 1996;67:413–422.

40 Mensah-Nyagan AG, Feuilloley M, Do-Rego JL, Marcual A, Lange C, Tonon MC, Pelletier G, Vaudry H: Localization of 17beta-hydroxysteroid dehydrogenase and characterization of testosterone in the brain of the male frog. Proc Natl Acad Sci USA 1996;93:1423–1428.

41 Patte-Mensah C, Kappes V, Freund-Mercier MJ, Tsutsui K, Mensah-Nyagan AG: Cellular distribution and bioactivity of the key steroidogenic enzyme, cytochrome P450side chain cleavage, in sensory neural pathways. J Neurochem 2003;86:1233–1246.

42 Patte-Mensah C, Li S, Mensah-Nyagan AG: Impact of neuropathic pain on the gene expression and activity of cytochrome P450 side-chain cleavage in sensory neural networks. Cell Mol Life Sci 2004;61:2274–2284.

43 Kibaly C, Patte-Mensah C, Mensah-Nyagan AG: Molecular and neurochemical evidence for the biosynthesis of dehydroepiandrosterone in the adult rat spinal cord. J Neurochem 2005;93:1220–1230.

44 Patte-Mensah C, Kibaly C, Mensah-Nyagan AG: Substance P inhibits progesterone conversion to neuroactive metabolites in spinal sensory circuit: a potential component of nociception. Proc Natl Acad Sci USA 2005;102:9044–9049.

45 Patte-Mensah C, Meyer L, Schaeffer V, Mensah-Nyagan AG: Selective regulation of 3 alpha-hydroxysteroid oxido-reductase expression in dorsal root ganglion neurons: a possible mechanism to cope with peripheral nerve injury-induced chronic pain. Pain 2010;150:522–534.

46 Scheuermann S, Hambsch B, Hesse L, Stumm J, Schmidt C, Beher D, Bayer TA, Beyreuther K, Multhaup G: Homodimerization of amyloid precursor protein and its implication in the amyloidogenic pathway of Alzheimer's disease. J Biol Chem 2001;276:33923–33929.

47 Schaeffer V, Patte-Mensah C, Eckert A, Mensah-Nyagan AG: Modulation of neurosteroid production in human neuroblastoma cells by Alzheimer's disease key proteins. J Neurobiol 2006;66:868–881.

48 Weingarten MD, Lockwood AH, Hwo SY, Kirschner MW: A protein factor essential for microtubule assembly. Proc Natl Acad Sci USA 1975;72:1858–1862.

49   Cleveland DW, Hwo SY, Kirschner MW: Purification of tau, a microtubule-associated protein that induces assembly of microtubules from purified tubulin. J Mol Biol 1977;116:207–225.

50   Cleveland DW, Hwo SY, Kirschner MW: Physical and chemical properties of purified tau factor and the role of tau in microtubule assembly. J Mol Biol 1977;116:227–247.

51   Drechsel DN, Hyman AA, Cobb MH, Kirschner MW: Modulation of the dynamic instability of tubulin assembly by the microtubule-associated protein tau. Mol Biol Cell 1992;3:1141–1154.

52   Brandt R, Lee G: Orientation, assembly, and stability of microtubule bundles induced by a fragment of tau protein. Cell Motil Cytoskeleton 1994;28:143–154.

53   Lewis J, McGowan E, Rockwood J, Melrose H, Nacharaju P, Van Slegtenhorst M, Gwinn-Hardy K, Paul Murphy M, Baker M, Yu X, Duff K, Hardy J, Corral A, Lin WL, Yen SH, Dickson DW, Davies P, Hutton M: Neurofibrillary tangles, amyotrophy and progressive motor disturbance in mice expressing mutant (P301L) tau protein. Nat Genet 2000;25:402–405.

54   Götz J, Chen F, Barmettler R, Nitsch RM: Tau filament formation in transgenic mice expressing P301L tau. J Biol Chem 2001;276:529–534.

55   Götz J, Chen F, van Dorpe J, Nitsch RM: Formation of neurofibrillary tangles in P301l tau transgenic mice induced by Abeta 42 fibrils. Science 2001;293:1491–1495.

56   David DC, Hauptmann S, Scherping I, Schuessel K, Keil U, Rizzu P, Ravid R, Drose S, Brandt U, Muller WE, Eckert A, Gotz J: Proteomic and functional analyses reveal a mitochondrial dysfunction in P301L tau transgenic mice. J Biol Chem 2005;280:23802–23814.

57   Murakami K, Fellous A, Baulieu EE, Robel P: Pregnenolone binds to microtubule-associated protein 2 and stimulates microtubule assembly. Proc Natl Acad Sci USA 2000;97:3579–3584.

58   Plassart-Schiess E, Baulieu EE: Neurosteroids: recent findings. Brain Res Brain Res Rev 2001;37:133–140.

59   Frye CA, McCormick CM: The neurosteroid, 3alpha-androstanediol, prevents inhibitory avoidance deficits and pyknotic cells in the granule layer of the dentate gyrus induced by adrenalectomy in rats. Brain Res 2000;855:166–170.

60   Reddy DS: Testosterone modulation of seizure susceptibility is mediated by neurosteroids 3alpha-androstanediol and 17beta-estradiol. Neuroscience 2004;129:195–207.

61   Schaeffer V, Meyer L, Patte-Mensah C, Eckert A, Mensah-Nyagan AG: Dose-dependent and sequence-sensitive effects of amyloid-beta peptide on neurosteroidogenesis in human neuroblastoma cells. Neurochem Int 2008;52:948–955.

Prof. Ayikoe G. Mensah-Nyagan
Equipe Stéroïdes, Neuromodulateurs et Neuropathologies, EA-4438
Université de Strasbourg, Bâtiment 3 de la Faculté de Médecine
11, rue Humann
FR–67 000 Strasbourg (France)
Tel. +33 368 85 3124, E-Mail gmensah@unistra.fr

Piguet P, Poindron P (eds): Genetically Modified Organisms and Genetic Engineering in Research and Therapy. BioValley Monogr. Basel, Karger, 2012, vol 3, pp 60–75

Investigating Therapeutic Strategies for Fragile X Syndrome in Mice

Aubin Michalon · Lothar Lindemann

Discovery Neuroscience, Pharmaceuticals Division, F. Hoffmann-La Roche Ltd., Basel, Switzerland

Abstract

Fragile X syndrome (FXS) is the most common form of inherited intellectual disability and a leading cause of autism. The identification of the mutation underlying the disease two decades ago has allowed the generation of a mouse model, the *Fmr1* knock-out, which presents behavioral, anatomical and biochemical alterations similar to the ones observed in the human condition. The rodent disease model with strong construct and face validity has been essential for our current understanding of the pathophysiology and remains an indispensable tool for testing therapeutic hypotheses. Two interventions targeting the metabotropic glutamate receptor 5 (mGlu5) and the GABA-B receptor, respectively, have been reported to correct a broad range of FXS phenotypes in *Fmr1* knockout mice. It is of great interest how these findings obtained from the FXS mouse model may translate into ongoing clinical research testing mGlu5 inhibitors and a GABA-B agonist in patients.

Fragile X syndrome (FXS) is the most common form of inherited mental retardation and the most frequent monogenic cause of autism. The condition is typically diagnosed at an age of 2–3 years or older, and is associated with a complex neuropsychiatric phenotype including deficits in learning and memory performance and cognitive capability, high anxiety and emotional instability, impaired social interaction and autistic behaviors, as well as susceptibility to seizures. The disease severity varies due to the variability of the underlying mutation, and males are generally more severely affected than females due to the linkage of the mutation to the X chromosome. FXS patients show a mild physical phenotype and have a normal life expectancy. In the absence of approved drugs, the current pharmacotherapy of FXS is typically based on a cocktail of psychiatric medications including anxiolytics, selective serotonin reuptake inhibitors (SSRI), benzodiazepines, antipsychotics, and anticonvulsants used off-label. For a long time it was assumed that genetically caused neurodevelopmental disorder such as FXS could only be treated symptomatically. For FXS, this view has witnessed a paradigm shift with recent studies in *Fmr1* knock-out mice reporting a correction of core behavioral and molecular phenotypes, suggesting the possibility of a disease-modifying potential of pharmacological interventions for FXS.

FXS Definition and Symptoms

FXS is an X-linked monogenic disorder which affects 1/4,000 boys on average and is half as frequent in girls. Typically, affected children exhibit a significant delay in language acquisition, poor social interaction and unusual sensorimotor development [1, 2]. Mental retardation is the predominant and most disabling problem, with an average IQ of about 40 in adult male patients and 80 in females, even though the severity of the symptoms is highly variable depending on genetic and environmental factors [3]. Difficulties with social interactions include shyness, poor eye contact and pronounced social anxiety. In fact, about one third of male FXS patients fulfill the diagnostic criteria for autism, which makes FXS the most frequent known genetic cause for autism. Epilepsy occurs in 10–40% of the cases during childhood and tends to resolve spontaneously during adolescence and early adulthood [4, 5]. In addition to the neuropsychiatric phenotype, patients also present some dysmorphic features including elongated ears and face, high arched palate, hyperflexibility of the joints, soft skin, flat feet and macroorchidism [6]. This reflects the fact that the mutated gene plays an important role not only in the brain but also in peripheral tissues.

Therapeutic options are currently limited to symptom management. In young boys, attention deficit, hyperactivity, impulsivity and overarousal are frequently managed with psychostimulants, whereas the use of antidepressants, particularly selective serotonin reuptake inhibitors, is more frequent in girls and adult patients to control anxiety, mood lability and perseverative, obsessive-compulsive behaviors [7]. Antipsychotics are also used when aberrant and aggressive behaviors, as well as self-directed injuries predominate. These medications are supportive and help minimize dysfunctional behaviors, but they do not target the cognitive deficits which are considered the core of the disease. In addition to pharmacological treatments, behavioral therapy is used with some success.

Genetic of FXS

The genetic cause of FXS has remained elusive for a long time due to the unusual inheritance pattern of the disease. The progressive worsening of symptoms across generations of affected females, and the fact that unaffected fathers could

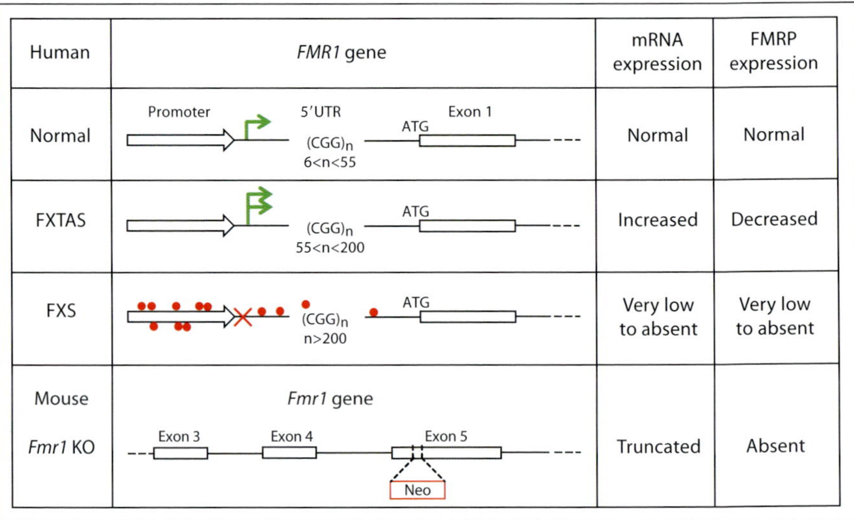

Fig. 1. Genetic of fragile X syndrome. FXS is caused by the expansion of a CGG repeat element within the 5′ untranslated region (UTR) of the *FMR1* gene, leading to hypermethylation of the gene (red dots) and transcription silencing. Premutation leading to fragile X tremor ataxia syndrome (FXTAS) is a condition caused by an intermediate CGG repeat number and a different disease mechanism with disease onset typically during the 5th or 6th decade of life. In the *Fmr1* KO mice [32], a neomycin resistance cassette (Neo) inserted within the exon 5 of the *Fmr1* gene leads to premature transcription stop and absence of the FMRP protein.

transmit the mutation was known as the Sherman paradox [8]. Following the identification of the mutation in 1991, it became clear that FXS results from the expansion of a CGG trinucleotide repeat within the 5′ untranslated region of the *FMR1* gene [9, 10]. In healthy individuals, the size of CGG repeats between 5 and 55 repeats, corresponds with allelic stability and absence of repeat expansion from one generation to the next. In contrast, alleles with more than 55 repeats are susceptible to meiotic and somatic instability leading to progressive repeat expansion across generations [3]. FXS is caused by a CGG repeat expansion to more than 200 repeats triggering hypermethylation and transcription silencing of the gene and ultimately to the lack of the gene product Fragile X Mental Retardation Protein (FMRP) (fig. 1). The intermediate range of 55 and 200 repeats called premutation is associated with the risk of developing Fragile X Tremor Ataxia syndrome (FXTAS) during the 5th and 6th decades of life. FXTAS is a neurodegenerative condition characterized by gait ataxia, intention tremor, and brain atrophy with intranuclear inclusions found in cortical and subcortical structures [11]. The premutation allele has been identified as strong genetic risk factor for premature ovarian failure [12]. The pathophysiology underlying FXTAS and premature ovarian failure differ fundamentally from FXS and will therefore not be discussed in the current review.

A pronounced genetic variability based on a combination of several factors contributes to a wide range of symptom severity. Repeat expansion takes place during meiotic and somatic cell division, causing different CGG repeat length occurring in the same individual. The transcriptional silencing of the *FMR1* gene is a function of the CGG repeat number, with >200 repeats defined as diagnostic criterion for FXS. However, the gene silencing is not a linear function of the CGG repeat number and differs between cells and individuals. Thus, partial gene silencing is observed already in patients with less than 200 repeats, and significant FMRP expression can still be observed in patients with more than 200 repeats. Finally, X chromosome inactivation in females leads to a mixture of cells with active X chromosomes harboring either the normal or mutated *FMR1* allele. Taken together, symptom severity is influenced by the combined effects of the CGG repeat number and residual FMRP expression levels [13, 14], as well as by the mosaicism of female X chromosome inactivation and somatic mutations in the *FMR1* gene.

FMRP Protein

FMRP is expressed in multiple tissues with high expression levels reported for brain, testis, ovary, thymus, eye, spleen and esophageal epithelium, and with moderate expression in colon, uterus, thyroid and liver [15, 16]. In contrast, no expression of FMRP occurs in the heart, aorta or muscle. FMRP expression in the brain is restricted to neurons where it is localized in the cytoplasm, dendrites and spines [17]; FMRP is absent from mature glia cells [18]. FMRP is observed in axons in the shape of discrete granules localized in a subset of brain regions including frontal cortex, the hippocampal CA3 area and olfactory bulb glomeruli [19]. The expression of FMRP is developmentally regulated with peak expression levels in mouse brain at the end of the first postnatal week, and stable, moderate FMRP expression levels in adulthood.

FMRP is an RNA-binding protein that is found almost exclusively in association with polyribosomes. FMRP functions as a repressor of mRNA translation which is regulated in an activity-dependent manner [20, 21]. Pulse-chase experiments with acute hippocampal brain slices from *Fmr1* knockout mice reveal an elevated protein synthesis rate by 15–20% caused by the absence of FMRP [22]. Initial in vitro experiments suggested that FMRP binds up to 4% of all mRNAs present in the brain [23, 24]. More recently, the use of crosslinking immunoprecipitation combined with high-throughput sequencing identified more than 600 mRNAs interacting directly with FMRP in vivo [25]. The corresponding proteins are implicated in various presynaptic and postsynaptic signaling pathways, such as synaptic long term potentiation and glutamate receptor signaling. The FMRP-mediated control of protein expression has been demonstrated for several proteins which are critical for synaptic activity such as the scaffolding proteins PSD95, Shank1 and MAP1B, the signaling proteins CaMK2alpha, and the ion channels NR1, NR2B, GluR1, Slack and Kv3.1b [26–31].

Mouse Model for FXS

The FXS mouse model was constructed by the targeted insertion of a neomycin cassette into exon 5 of the *Fmr1* gene leading to a complete lack of FMRP expression throughout the organism [32] (fig. 1). This approach avoids the genetic variability associated with the trinucleotide expansion, thereby providing a more consistent phenotype and facilitating the use of the mice as test system for genetic and pharmacological interventions. A transgenic mouse line with an extended CGG repeat in the 5′-UTR of the mouse *Fmr1* gene has been described as a model for FXTAS [33]. This mouse line shows CGG repeat expansion across generations but no corresponding increase in the methylation of the *Fmr1* gene and only moderate reduction in FMRP expression even with more than 200 repeats. A conditional *Fmr1* knock-out allowing the Cre recombinase-driven deletion of FMRP expression has been generated [34] but has not been used to date to investigate FMRP function in different cell types. The characterization of the *Fmr1* KO mouse phenotype has been challenging because core phenotypes, e.g. relating to the learning and memory deficits of the mice are fairly modest, and because no clear deficits were observed initially in the most classical, hippocampus-dependent tests such as Morris water maze or contextual fear conditioning [32, 35, 36]. In addition, the genetic background, environmental conditions in which the animals were raised and handled, and the exact procedures of the tests have a strong influence on most of the known phenotypes [37, 38].

The *Fmr1* KO mice exhibit deficits in several domains related to the core symptoms of FXS, including learning and memory deficit, altered social behavior, altered response to sensory stimuli, and exacerbated sensitivity to epilepsy (table 1). Learning and memory impairments were observed in the inhibitory avoidance task (passive avoidance) and lever press escape/avoidance task (active avoidance) in several independent studies [39–42]. Learning deficits were also observed in the trace fear conditioning paradigm [43], as well as in the reversal learning in the water maze paradigm [32, 36, 44, 45]. Alterations in social behavior were observed in different tests, collectively suggesting elevated social anxiety in *Fmr1* KO [46–48]. *Fmr1* KO mice show an elevated whole body startle in response to low intensity stimuli (<90 dB) [49], reflecting the hypersensitivity to sensory stimuli reported in FXS patients [6, 50]. The elevated sensitivity to audiogenic seizures is the most robust phenotype in *Fmr1* KO mice which was reported in multiple studies [41, 51–53]. It reproduces the high prevalence of epilepsy in FXS patients with estimates varying from 14 to 44% of patients affected depending on the study [4, 54].

Table 1. Core symptoms of FXS are reproduced in *Fmr1* KO mice

FXS patients	*Fmr1* KO mouse
Intellectual disability	Learning and memory deficits
Atypical social development with high social anxiety	Alterations in social interactions
Epilepsy in children	High susceptibility to audiogenic seizures
Elevated spine density	Elevated spine density
Immature spine shape	Long and thin spines
Macroorchidism	Macroorchidism

Another important phenotypic similarity between FXS patients and *Fmr1* KO mice relates to the dendritic spine morphology. The analysis of spine density and morphology in postmortem cortical tissue from FXS patients revealed an elevated density of dendritic spines in FXS patients compared to age-matched controls [55, 56]. In addition, an increase fraction of immature spines with a thin and long, filopodia-like shape was reported. A similar phenotype was observed in the brain of *Fmr1* KO mice, in the visual, primary somatosensory cortex and occipital cortices [57, 58].

With respect to synaptic plasticity, long-term potentiation (LTP) of CA1 synapses appeared to be normal when induced with tetanic stimulation in the Schaffer collaterals [35, 59, 60], but a deficit in hippocampal LTP could be detected at lower levels of stimulation using theta bursts [61], which are more similar to endogenous neuronal activity than tetanic stimulations. Similarly, the intrinsic firing properties of layer 5 pyramidal neurons in the somatosensory cortex as well as the synaptic potentiation in response to tetanic stimulation are intact in juvenile *Fmr1* KO mice, which however exhibit a specific deficit in spike-timing dependent potentiation [62]. Alterations in synaptic plasticity are certainly region-specific and dependent on the age of the animals, because cortical LTP in response to tetanic stimulation is impaired in adult *Fmr1* KO mice [43, 59]. A specific form of long-term depression (LTD) in the hippocampus induced by brief application of the mGlu1/5 agonist DHPG is exaggerated and protein synthesis-independent in *Fmr1* KO in contrast to wild-type mice [63, 64]. In addition to cortex and hippocampus, impaired synaptic plasticity has also been shown in the amygdala [65, 66] and the cerebellum of *Fmr1* KO mice [67]. Taken together, these results indicate that the potential to induce LTP is weakened in *Fmr1* KO mice compared to WT mice, whereas LTD is more pronounced in the KO mice.

Two directions of research which have emerged from studies with the *Fmr1* KO mouse model will be discussed below: One area of research based on the so-called 'mGluR theory of Fragile X' [68] investigates the therapeutic use of mGlu5 inhibitors for FXS patients. Another area of research is based on the finding that inhibitors of GABA receptors can correct several phenotypes of *Fmr1* KO mice.

Altered Glutamatergic Signaling in FXS

As mentioned earlier, FMRP is a repressor of mRNA translation and, in absence of FMRP, the rate of protein synthesis is elevated in multiple brain areas [69]. Various forms of long-term synaptic plasticity are associated with precise regulation of protein synthesis in dendrites. MGlu1/5-LTD requires local protein synthesis [70]; this specific form of synaptic plasticity is elevated in the hippocampus of *Fmr1* KO mice and, in contrast to WT animals, is independent of protein synthesis [63]. MGlu1/5-LTD is mediated by internalization of AMPA receptors and leads to elongation and thinning of the spines via a protein synthesis-dependent mechanism [71]. The elevated rate of protein synthesis in *Fmr1* KO mice is associated with reduced surface expression of AMPA receptors and a high density of long and thin spines [29, 59]. Long-lasting activation of mGlu1/5 receptors in hippocampal slices leads to epileptiform bursts in the CA3 area via a protein synthesis dependent mechanism, whereas epileptiform activity occurs spontaneously in *Fmr1* KO mice [72]. The apparent similarities between the consequences of mGlu1/5 activation and FXS phenotypes suggested that excessive mGlu1/5 activity may play a causal role in the disease mechanism. These observations were condensed in the mGlu theory of FXS suggesting that inhibition of mGlu1/5 receptors may provide therapeutic benefits for FXS [68].

The mGlu theory of FXS gained strong support from an elegant study using a genetic approach to reduce mGlu5 expression level by 50% which was achieved by an intercross between *Fmr1* KO and *Grm5* (mGlu5 gene) KO mice, resulting in mice with complete *Fmr1* deletion and a heterozygous *Grm5* allele. Reducing mGlu5 expression by 50% on the *Fmr1* KO background completely prevented the occurrence of numerous key *Fmr1* KO phenotypes including the learning deficit, altered ocular-dominance plasticity, elevated spine density and elevated mGlu1/5-LTD [41]. Several pharmacological studies have reported a correction of hypersensitivity to audiogenic seizures and elevated prepulse inhibition [73], as well as the spine phenotype on primary neuron cultures in vitro by acute treatment with mGlu5 antagonists such as MPEP, MTEP or Fenobam [74–76], but the limited half-life of these compounds in vivo made chronic experiments very challenging. Chronic pharmacological mGlu5 inhibition in vivo has been facilitated with the recent discovery of CTEP, a new potent and selective, long acting and orally bioavailable mGlu5 inhibitor [77]. Chronic treatment of *Fmr1* KO mice with CTEP starting in early adulthood, i.e. after symptom onset, could fully correct the core cognitive deficits of FXS in mice, as well as the elevated protein synthesis rate, mGlu1/5-LTD, the aberrant spine density and the altered intracellular signaling [53]. These new results suggest that a pharmacological correction of key FXS phenotypes can be achieved with treatment starting after phenotype onset, and that mGlu5 inhibitors might have disease modifying potential in FXS (table 2).

Several mGlu5 inhibitors including RO4917523 (F. Hoffmann-La Roche Ltd.), AFQ056 (Novartis), and STX107 (Seaside Therapeutics) are currently being tested in FXS patients. A short duration study with the mGlu5 inhibitor AFQ056 in 30 male adult FXS patients reported significant behavioral effects in a subgroup of patients [78]. Further clinical studies with longer treatment duration targeting a broad age range are under way. It will be of great interest if mGlu5 inhibitors can address the clinical phenotypes in a similar broad fashion and with similar magnitude as suggested by preclinical data.

Altered GABAergic Signaling in FXS

FMRP is also expressed in GABAergic neurons, suggesting altered GABA signaling in FXS.

The expression levels of several GABA-A subunits are moderately reduced in *Fmr1* KO compared to WT mice [79–81]. The altered GABA-A subunit expression levels are specific to certain brain regions and the age of the animals. For example, the expression of the α1 subunit in the cortex of *Fmr1* KO mice is reduced 5 days after birth but normal in adult animals, whereas the expression of the ß2 subunit is normal in young animals and reduced in adults [82]. In the subiculum of adult *Fmr1* KO animals, the expression level of the α5 and δ subunits are decreased which is associated with a decreased level of tonic GABAergic inhibition of pyramidal cells [83]. In the striatum, the density of GABAergic synapses is decreased but spontaneous GABAergic activity is increased, suggesting a disinhibition of GABAergic neurons [84]. In the somatosensory cortex but not in the hippocampus of *Fmr1* KO mice, the density of parvalbumin-positive interneurons is decreased by 20% and their distribution in the different cortical layers is altered. In contrast, there is no difference in the density and lamina distribution of calbindin- and calretinin-positive interneurons between *Fmr1* KO and WT mice. In the amygdala, the frequency and amplitude of spontaneous inhibitory post-synaptic currents is decreased by 50%, reflecting diminished inhibitory transmission in *Fmr1* KO mice [85]. The number of inhibitory synapses and the expression level of GAD65/67, the rate-limiting enzyme for GABA synthesis, are reduced in the amygdala [85]. Finally, a study of parvalbumin-expressing fast-spiking interneurons in primary somatosensory (barrel) cortex revealed that inhibitory transmission onto excitatory neurons is normal in *Fmr1* KO mice, but that the excitatory input to inhibitory neurons is substantially reduced [86]. The reduced activation of GABAergic neurons leads to an increased activity of excitatory neurons

Table 2. Genetic and pharmacological correction of FXS in mice

Fmr1 KO phenotype	*Fmr1* KO/*Grm5*[+/−]	*Fmr1* KO / mGlu5 antagonist
Elevated protein synthesis rate	√	√
Elevated mGlu-LTD	√	√
Elevated dendritic spine density	√	√
Altered ocular dominance plasticity	√	n.a.
Impaired inhibitory avoidance	√	√
Hypersensitivity to auditory stimuli	n.a.	√
Sensitivity to audiogenic seizures	√	√
Altered ERK1,2 and mTOR signaling	n.a.	√
Macroorchidism	x	p.c.

The genetic or pharmacological reduction of mGlu5 activity reverses the biochemical, electro-physiological and behavioral deficits in adult *Fmr1* KO mice [41, 53]. √ = Complete correction; p.c. = partial correction, x = no effect, n.a. = not available.

and a corresponding increase in network activity (UP states) associated with a decrease in local network synchronization. This diminished local inhibitory feedback loop leads to hyperexcitability of neuronal circuits in the somatosensory cortex and may explain the hypersensitivity of FXS patients to sensory stimuli [50], as well as the high frequency of epilepsy [4]. Taken together, these studies reveal a substantial hypoactivity of GABAergic neurotransmission and an imbalance between inhibitory and excitatory neuronal activity in FXS. This suggests that pharmacological activation of the GABAergic system may have a therapeutic potential in FXS, an idea that has been extensively discussed elsewhere [87–89].

The therapeutic potential for GABAergic system activation has been explored in mice using gaboxadol, an agonist for GABA-A receptors containing δ subunits [90]. In vitro, bath application of gaboxadol could rescue the hyperexcitability phenotype of principal neurons in the basolateral amygdala in *Fmr1* KO mice [85]. In vivo, acute application of gaboxadol could rescue the elevated locomotor activity in *Fmr1* KO mice at a dose not affecting locomotion in WT mice; the same dose of gaboxadol did not ameliorate the learning deficit in the fear conditioning task nor the elevated PPI in *Fmr1* KO mice [91].

Another approach targeting the GABA system in FXS has employed baclofen, a drug approved for the treatment of spastic movement disorders, especially in instances of spinal cord injury, cerebral palsy, and multiple sclerosis. Baclofen is an agonist of the GABA-B receptors, which activation results in neuron hyperpolarization and inhibition of presynaptic release at glutamatergic synapses. GABA-B activation would either compensate for increased excitatory glutamatergic signaling or boost intrinsically deficient inhibitory signaling. In *Fmr1* KO mice, baclofen effectively suppresses the elevated susceptibility to audiogenic seizures [92]. Reports from a recent clinical trial testing R-baclofen, the active enantiomer of baclofen, in adolescent and adult FXS patients suggest positive effects in multiple behavioral domains [Seaside Therapeutics press release 2010]. Further clinical studies investigating R-baclofen in FXS are ongoing.

Conclusions

The last two decades have substantially advanced our understanding of the FXS disease mechanism. The identification of the mutation in 1991, and the early availability of FXS disease models – most notably the *Fmr1* KO mouse – have been instrumental for exploring pharmacological treatments. The most thoroughly investigated interventions for FXS are mGlu5 inhibitors and the GABA-B agonist R-baclofen. The progress summarized here for FXS has a broader relevance, representing a paradigm shift in the view of monogenetic autism spectrum disorders which until recently were considered essentially not treatable.

Other examples for the pharmacological treatment of monogenetic autism spectrum disorders have been reported for tuberous sclerosis, Rett syndrome, and neurofibromatosis. Tuberous sclerosis (TSC) is a single-gene disorder caused by heterozygous mutations in the TSC1 or TSC2 genes and is frequently associated with mental retardation, autism and epilepsy. The hyperactivation of mTOR observed in the disease condition can be corrected by administration of rapamycin, a specific mTOR inhibitor, and treatment of $Tsc2^{+/-}$ mice with the nonselective mTOR inhibitor rapamycin for 5 days corrects the spatial learning deficit in adult mice [93]. Rett syndrome is another example of monogenic disorder with autistic features, which results from mutations in the MeCP2 gene. The reversibility of the phenotype was initially demonstrated with a genetic approach allowing re-expression of the MeCP2 gene in adult animals [94], and it was reported more recently that systemic treatment of MeCP2 mutant mice with an active peptide fragment of insulin-like growth factor (IGF-1) extended the life span of the mice, improved locomotor function and ameliorated breathing patterns and irregular heart rate [95]. Lastly, neurofibromatosis 1, which results from mutations in the NF1 gene, manifests by nerve tissue tumors called neurofibromas and is associated with a high frequency of autism traits and intellectual disability. In adult mice, the learning deficits resulting from elevated Ras activity can be corrected by treatment with farnesyl-transferase inhibitors [96–98].

Taken together, the preclinical data summarized for FXS demonstrate that pharmacological treatment during adulthood can reverse the biochemical, electrophysiological and behavioral deficits observed in FXS. Data available for the currently best studied pharmacological interventions for FXS – mGlu5 inhibitors and R-baclofen – suggest the possibility of a pharmacological therapy beyond symptomatic treatment with disease-modifying potential. FXS represents a success story for translating the understanding of the molecular disease mechanism gained through a mouse model faithfully reflecting behavioral and biochemical features of the human disease into advanced clinical research. The case of FXS was critical for changing our view on monogenetic developmental disorders for which a meaningful pharmacological treatment is now considered a realistic possibility. This paradigm shift has inspired research on other monogenetic disorders such as tuberous sclerosis, Rett syndrome and neurofibromatosis. Mouse models with strong face and construct validity will continue to be instrumental for further advancing our understanding of these diseases and their possible treatment.

References

1   Berry-Kravis E, Grossman AW, Crnicc LS, Greenough WT: Understanding fragile X syndrome. Current Pediatrics 2002;12:316–324.
2   Hagerman RJ, Hagerman PJ: Fragile X Syndrome: Diagnosis, Treatment and Research. Baltimore, The John Hopkins University Press, 2002.
3   Rousseau F, Heitz D, Tarleton J, MacPherson J, Malmgren H, Dahl N, Barnicoat A, Mathew C, Mornet E, Tejada I, et al: A multi-center study on genotype-phenotype correlations in the fragile X syndrome, using direct diagnosis with probe StB12.3: the first 2,253 cases. Am J Hum Genet 1994;55:225–237.
4   Berry-Kravis E: Epilepsy in fragile X syndrome. Dev Med Child Neurol 2002;44:724–728.
5   Hagerman PJ, Stafstrom CE: Origins of epilepsy in fragile X syndrome. Epilepsy Currents/Am Epilepsy Soc 2009;9:108–112.
6   Hagerman RJ: Physical and behavioral phenotype; in Hagerman RJ, Hagerman PJ (eds): Fragile X Syndrome: Diagnosis, Treatment and Research. Baltimore, The John Hopkins University Press, 2002.
7   Berry-Kravis E, Potanos K: Psychopharmacology in fragile X syndrome – present and future. Mental Retard Dev Disab Res Rev 2004;10:42–48.
8   Sherman SL, Jacobs PA, Morton NE, Froster-Iskenius U, Howard-Peebles PN, Nielsen KB, Partington MW, Sutherland GR, Turner G, Watson M: Further segregation analysis of the fragile X syndrome with special reference to transmitting males. Hum Genet 1985;69:289–299.
9   Fu YH, Kuhl DP, Pizzuti A, Pieretti M, Sutcliffe JS, Richards S, Verkerk AJ, Holden JJ, Fenwick RG Jr, Warren ST, et al: Variation of the CGG repeat at the fragile X site results in genetic instability: resolution of the Sherman paradox. Cell 1991;67:1047–1058.
10 Verkerk AJ, Pieretti M, Sutcliffe JS, Fu YH, Kuhl DP, Pizzuti A, Reiner O, Richards S, Victoria MF, Zhang FP, et al: Identification of a gene (FMR-1) containing a CGG repeat coincident with a breakpoint cluster region exhibiting length variation in fragile X syndrome. Cell 1991;65:905–914.
11 Hagerman PJ, Hagerman RJ: Fragile X-associated tremor/ataxia syndrome (FXTAS). Mental Retard Dev Disab Res Rev 2004;10:25–30.
12 Sullivan AK, Marcus M, Epstein MP, Allen EG, Anido AE, Paquin JJ, Yadav-Shah M, Sherman SL: Association of FMR1 repeat size with ovarian dysfunction. Hum Reprod (Oxford) 2005;20:402–412.

13   Tassone F, Hagerman RJ, Ikle DN, Dyer PN, Lampe M, Willemsen R, Oostra BA, Taylor AK: FMRP expression as a potential prognostic indicator in fragile X syndrome. Am J Med Genet 1999;84:250–261.

14   Loesch DZ, Bui QM, Dissanayake C, Clifford S, Gould E, Bulhak-Paterson D, Tassone F, Taylor AK, Hessl D, Hagerman R, Huggins RM: Molecular and cognitive predictors of the continuum of autistic behaviours in fragile X. Neurosci Biobehav Rev 2007;31:315–326.

15   Devys D, Lutz Y, Rouyer N, Bellocq JP, Mandel JL: The FMR-1 protein is cytoplasmic, most abundant in neurons and appears normal in carriers of a fragile X premutation. Nat Genet 1993;4:335–340.

16   Hinds HL, Ashley CT, Sutcliffe JS, Nelson DL, Warren ST, Housman DE, Schalling M: Tissue specific expression of FMR-1 provides evidence for a functional role in fragile X syndrome. Nat Genet 1993;3:36–43.

17   Feng Y, Gutekunst CA, Eberhart DE, Yi H, Warren ST, Hersch SM: Fragile X mental retardation protein: nucleocytoplasmic shuttling and association with somatodendritic ribosomes. J Neurosci 1997;17:1539–1547.

18   Pacey LK, Doering LC: Developmental expression of FMRP in the astrocyte lineage: implications for fragile X syndrome. Glia 2007;55:1601–1609.

19   Christie SB, Akins MR, Schwob JE, Fallon JR: The FXG: a presynaptic fragile X granule expressed in a subset of developing brain circuits. J Neurosci 2009;29:1514–1524.

20   Bassell GJ, Warren ST: Fragile X syndrome: loss of local mRNA regulation alters synaptic development and function. Neuron 2008;60:201–214.

21   Liu-Yesucevitz L, Bassell GJ, Gitler AD, Hart AC, Klann E, Richter JD, Warren ST, Wolozin B: Local RNA translation at the synapse and in disease. J Neurosci 2011;31:16086–16093.

22   Osterweil EK, Krueger DD, Reinhold K, Bear MF: Hypersensitivity to mGluR5 and ERK1/2 leads to excessive protein synthesis in the hippocampus of a mouse model of fragile X syndrome. J Neurosci 2010;30:15616–15627.

23   Brown V, Jin P, Ceman S, Darnell JC, O'Donnell WT, Tenenbaum SA, Jin X, Feng Y, Wilkinson KD, Keene JD, Darnell RB, Warren ST: Microarray identification of FMRP-associated brain mRNAs and altered mRNA translational profiles in fragile X syndrome. Cell 2001;107:477–487.

24   Miyashiro KY, Beckel-Mitchener A, Purk TP, Becker KG, Barret T, Liu L, Carbonetto S, Weiler IJ, Greenough WT, Eberwine J: RNA cargoes associating with FMRP reveal deficits in cellular functioning in Fmr1 null mice. Neuron 2003;37:417–431.

25   Darnell JC, Van Driesche SJ, Zhang C, Hung KY, Mele A, Fraser CE, Stone EF, Chen C, Fak JJ, Chi SW, Licatalosi DD, Richter JD, Darnell RB: FMRP stalls ribosomal translocation on mRNAs linked to synaptic function and autism. Cell 2011;146:247–261.

26   Lu R, Wang H, Liang Z, Ku L, O'Donnell WT, Li W, Warren ST, Feng Y: The fragile X protein controls microtubule-associated protein 1B translation and microtubule stability in brain neuron development. Proc Natl Acad Sci USA 2004;101:15201–15206.

27   Muddashetty RS, Kelic S, Gross C, Xu M, Bassell GJ: Dysregulated metabotropic glutamate receptor-dependent translation of AMPA receptor and postsynaptic density-95 mRNAs at synapses in a mouse model of fragile X syndrome. J Neurosci 2007;27:5338–5348.

28   Zalfa F, Eleuteri B, Dickson KS, Mercaldo V, De Rubeis S, di Penta A, Tabolacci E, Chiurazzi P, Neri G, Grant SG, Bagni C: A new function for the fragile X mental retardation protein in regulation of PSD-95 mRNA stability. Nat Neurosci 2007;10:578–587.

29   Schutt J, Falley K, Richter D, Kreienkamp HJ, Kindler S: Fragile X mental retardation protein regulates the levels of scaffold proteins and glutamate receptors in postsynaptic densities. J Biol Chem 2009;284:25479–25487.

30   Strumbos JG, Brown MR, Kronengold J, Polley DB, Kaczmarek LK: Fragile X mental retardation protein is required for rapid experience-dependent regulation of the potassium channel Kv3.1b. J Neurosci 2010;30:10263–10271.

31 Brown MR, Kronengold J, Gazula VR, Chen Y, Strumbos JG, Sigworth FJ, Navaratnam D, Kaczmarek LK: Fragile X mental retardation protein controls gating of the sodium-activated potassium channel Slack. Nat Neurosci 2010;13:819–821.

32 Bakker CE, Verheij C, Willemsen R, van der Helm R, Oerlemans F, Vermey M, Bygrave A, Hoogeveen T, Oostra BA, Reyniers E, De Boulle K, D'Hooge R, Cras P, van Velzen D, Nagels G, Martin J, De Deyn PP, Darby JK, Willems PJ: Fmr1 knockout mice: a model to study fragile X mental retardation. The Dutch-Belgian Fragile X Consortium. Cell 1994;78:23–33.

33 Brouwer JR, Huizer K, Severijnen LA, Hukema RK, Berman RF, Oostra BA, Willemsen R: CGG-repeat length and neuropathological and molecular correlates in a mouse model for fragile X-associated tremor/ataxia syndrome. J Neurochem 2008;107:1671–1682.

34 Mientjes EJ, Nieuwenhuizen I, Kirkpatrick L, Zu T, Hoogeveen-Westerveld M, Severijnen L, Rife M, Willemsen R, Nelson DL, Oostra BA: The generation of a conditional Fmr1 knock out mouse model to study Fmrp function in vivo. Neurobiol Dis 2006;21:549–555.

35 Paradee W, Melikian HE, Rasmussen DL, Kenneson A, Conn PJ, Warren ST: Fragile X mouse: strain effects of knockout phenotype and evidence suggesting deficient amygdala function. Neuroscience 1999;94:185–192.

36 Van Dam D, D'Hooge R, Hauben E, Reyniers E, Gantois I, Bakker CE, Oostra BA, Kooy RF, De Deyn PP: Spatial learning, contextual fear conditioning and conditioned emotional response in Fmr1 knockout mice. Behav Brain Res 2000;117:127–136.

37 Bernardet M, Crusio WE: Fmr1 KO mice as a possible model of autistic features. Sci World J 2006;6:1164–1176.

38 Bhogal B, Jongens TA: Fragile X syndrome and model organisms: identifying potential routes of therapeutic intervention. Dis Models Mech 2010;3:693–700.

39 Yuskaitis CJ, Mines MA, King MK, Sweatt JD, Miller CA, Jope RS: Lithium ameliorates altered glycogen synthase kinase-3 and behavior in a mouse model of fragile X syndrome. Biochem Pharmacol 2010;79:632–646.

40 Liu ZH, Chuang DM, Smith CB: Lithium ameliorates phenotypic deficits in a mouse model of fragile X syndrome. Int J Neuropsychopharmacol 2010:1–13.

41 Dolen G, Osterweil E, Rao BS, Smith GB, Auerbach BD, Chattarji S, Bear MF: Correction of fragile X syndrome in mice. Neuron 2007;56:955–962.

42 Brennan FX, Albeck DS, Paylor R: Fmr1 knockout mice are impaired in a leverpress escape/avoidance task. Genes Brain Behav 2006;5:467–471.

43 Zhao MG, Toyoda H, Ko SW, Ding HK, Wu LJ, Zhuo M: Deficits in trace fear memory and long-term potentiation in a mouse model for fragile X syndrome. J Neurosci 2005;25:7385–7392.

44 Kooy RF, D'Hooge R, Reyniers E, Bakker CE, Nagels G, De Boulle K, Storm K, Clincke G, De Deyn PP, Oostra BA, Willems PJ: Transgenic mouse model for the fragile X syndrome. Am J Med Genet 1996;64:241–245.

45 D'Hooge R, Nagels G, Franck F, Bakker CE, Reyniers E, Storm K, Kooy RF, Oostra BA, Willems PJ, De Deyn PP: Mildly impaired water maze performance in male Fmr1 knockout mice. Neuroscience 1997;76:367–376.

46 Spencer CM, Alekseyenko O, Serysheva E, Yuva-Paylor LA, Paylor R: Altered anxiety-related and social behaviors in the Fmr1 knockout mouse model of fragile X syndrome. Genes Brain Behav 2005;4:420–430.

47 Mineur YS, Huynh LX, Crusio WE: Social behavior deficits in the Fmr1 mutant mouse. Behav Brain Res 2006;168:172–175.

48 McNaughton CH, Moon J, Strawderman MS, Maclean KN, Evans J, Strupp BJ: Evidence for social anxiety and impaired social cognition in a mouse model of fragile X syndrome. Behav Neurosci 2008;122:293–300.

49 Chen L, Toth M: Fragile X mice develop sensory hyperreactivity to auditory stimuli. Neuroscience 2001;103:1043–1050.

50 Miller LJ, McIntosh DN, McGrath J, Shyu V, Lampe M, Taylor AK, Tassone F, Neitzel K, Stackhouse T, Hagerman RJ: Electrodermal responses to sensory stimuli in individuals with fragile X syndrome: a preliminary report. Am J Med Genet 1999;83:268–279.

51 Musumeci SA, Bosco P, Calabrese G, Bakker C, De Sarro GB, Elia M, Ferri R, Oostra BA: Audiogenic seizures susceptibility in transgenic mice with fragile X syndrome. Epilepsia 2000;41:19–23.

52 Yan QJ, Asafo-Adjei PK, Arnold HM, Brown RE, Bauchwitz RP: A phenotypic and molecular characterization of the fmr1-tm1Cgr fragile X mouse. Genes Brain Behav 2004;3:337–359.

53 Michalon A, Sidorov M, Ballard T, Ozmen L, Spooren W, Wettstein JG, Jaeschke G, Bear MF, Lindemann L: Chronic pharmacological mGlu5 inhibition corrects fragile X in adult mice. Neuron 2012;74:49–56.

54 Musumeci SA, Hagerman RJ, Ferri R, Bosco P, Dalla Bernardina B, Tassinari CA, De Sarro GB, Elia M: Epilepsy and EEG findings in males with fragile X syndrome. Epilepsia 1999;40:1092–1099.

55 Hinton VJ, Brown WT, Wisniewski K, Rudelli RD: Analysis of neocortex in three males with the fragile X syndrome. Am J Med Genet 1991;41:289–294.

56 Irwin SA, Patel B, Idupulapati M, Harris JB, Crisostomo RA, Larsen BP, Kooy F, Willems PJ, Cras P, Kozlowski PB, Swain RA, Weiler IJ, Greenough WT: Abnormal dendritic spine characteristics in the temporal and visual cortices of patients with fragile-X syndrome: a quantitative examination. Am J Med Genet 2001;98:161–167.

57 Galvez R, Greenough WT: Sequence of abnormal dendritic spine development in primary somatosensory cortex of a mouse model of the fragile X mental retardation syndrome. Am J Med Genet A 2005;135:155–160.

58 McKinney BC, Grossman AW, Elisseou NM, Greenough WT: Dendritic spine abnormalities in the occipital cortex of C57BL/6 Fmr1 knockout mice. Am J Med Genet B Neuropsychiatr Genet 2005;136B:98–102.

59 Li J, Pelletier MR, Perez Velazquez JL, Carlen PL: Reduced cortical synaptic plasticity and GluR1 expression associated with fragile X mental retardation protein deficiency. Mol Cell Neurosci 2002;19:138–151.

60 Godfraind JM, Reyniers E, De Boulle K, D'Hooge R, De Deyn PP, Bakker CE, Oostra BA, Kooy RF, Willems PJ: Long-term potentiation in the hippocampus of fragile X knockout mice. Am J Med Genet 1996;64:246–251.

61 Lauterborn JC, Rex CS, Kramar E, Chen LY, Pandyarajan V, Lynch G, Gall CM: Brain-derived neurotrophic factor rescues synaptic plasticity in a mouse model of fragile X syndrome. J Neurosci 2007;27:10685–10694.

62 Desai NS, Casimiro TM, Gruber SM, Vanderklish PW: Early postnatal plasticity in neocortex of Fmr1 knockout mice. J Neurophysiol 2006;96:1734–1745.

63 Huber KM, Gallagher SM, Warren ST, Bear MF: Altered synaptic plasticity in a mouse model of fragile X mental retardation. Proc Natl Acad Sci USA 2002;99:7746–7750.

64 Hou L, Antion MD, Hu D, Spencer CM, Paylor R, Klann E: Dynamic translational and proteasomal regulation of fragile X mental retardation protein controls mGluR-dependent long-term depression. Neuron 2006;51:441–454.

65 Suvrathan A, Hoeffer CA, Wong H, Klann E, Chattarji S: Characterization and reversal of synaptic defects in the amygdala in a mouse model of fragile X syndrome. Proc Natl Acad Sci USA 2010;107:11591–11596.

66 Suvrathan A, Chattarji S: Fragile X syndrome and the amygdala. Curr Opin Neurobiol 2011;21:509–515.

67 Huber KM: The fragile X-cerebellum connection. Trends Neurosci 2006;29:183–185.

68 Bear MF, Huber KM, Warren ST: The mGluR theory of fragile X mental retardation. Trends Neurosci 2004;27:370–377.

69 Qin M, Kang J, Burlin TV, Jiang C, Smith CB: Postadolescent changes in regional cerebral protein synthesis: an in vivo study in the FMR1 null mouse. J Neurosci 2005;25:5087–5095.

70 Huber KM, Kayser MS, Bear MF: Role for rapid dendritic protein synthesis in hippocampal mGluR-dependent long-term depression. Science 2000;288:1254–1257.

71 Gladding CM, Fitzjohn SM, Molnar E: Metabotropic glutamate receptor-mediated long-term depression: molecular mechanisms. Pharmacol Rev 2009;61:395–412.

Piguet P, Poindron P (eds): Genetically Modified Organisms and Genetic Engineering in Research and Therapy. BioValley Monogr. Basel, Karger, 2012, vol 3, pp 76–85

Innovative Therapeutic Perspectives in Neuromuscular Diseases

Susan Cure · Serge Braun

Association Française Contre les Myopathies, Evry, France

Abstract

Therapies of neuromuscular diseases must address a number of challenges that relate mainly to the tissues involved and the severity of the pathophenotypes. Numerous innovative strategies are being developed to tackle this heterogenous group of diseases, paving the way for broader applications. They may be subdivided into two groups: gene-based, and cytoprotective/regenerative apporoaches. Considerable progress has been made in gene therapy, and RNA and DNA 'surgery' targeting both the muscle and the peripheral nervous system. Some of them are about to reach the market. Pharmacologically- or cell-induced tissue regeneration have matured and could be considered alone or in combination with the above strategies and with pharmacological treatments aiming at preventing/reducing the degeneration/degradation of neuronal, muscular or cardiac cells.   Copyright © 2012 S. Karger AG, Basel

The treatment of neuromuscular diseases faces numerous challenges: (a) the need to correct and preserve the totality of the target tissue (and in particular muscle, which represents half of the total mass of the organism) in a durable manner, (b) serious disorders beginning in infancy and evolving towards a progressive disappearance of functional tissue, and (c) occurrence of numerous de novo mutations (1/3 of cases in Duchenne muscular dystrophy – DMD) which cause a considerable variety of genetic anomalies which are unique for each patient [1]. Therapeutic strategies may be classified as therapies based on knowledge of the genes, and cytoprotective and regenerative strategies.

Therapies Based on Our Knowledge of the Genes

Gene Transfer

Gene therapy comes in many forms: transfer of the complete or partial coding sequence of the healthy gene, RNA or DNA surgery, transfer or induction of other genes with a therapeutic goal. The first clinical trial attempted was for DMD. It consisted in administering the complete coding sequence of the dystrophin gene, carried by a nonviral vector, plasmid DNA [2]. In the phase I trial carried out in 2001–2003, a low dose of vector was used (600 µg) leading to very low and local expression of dystrophin (along the trajectory of the needle); without the ability to completely exclude an immune response in case of a massive dose of the vector [3]. Hydrodynamic limb vein (HLV) administration of the plasmid to the limbs – which was developed in primate, rodent (*mdx* mice) and canine (GRMD dog) models – was shown to be efficient and well tolerated. Transfection efficacy does not seem to depend on the age of the animals, and it persists over the long term (virtually the life span of mice, and many years in monkeys) [4]. It is possible to deliver large quantities of plasmid (up to several hundred mg) to the base of the limb by the intravenous route with the help of an inflatable tourniquet, thus limiting the zone to be treated [5]. All patients could benefit from this approach [6]. Efficacy of transfection in the primate is quite good (40% of the fibers in some muscles), but could be improved. This is possible thanks to viral vectors such as adeno-associated viruses (AAV). These viruses cannot accommodate the complete coding sequence of dystrophin; shortened versions of the gene have been generated which are capable of correcting the dystrophic phenotype of *mdx* mouse [7] and GRMD dog models. A clinical trial for safety (phase 1a, randomized double-blind) using intramuscular administration of an AAV2.5-minidystrophin (Biodystrophin®) is in progress in patients aged 5–11 years. No serious adverse events have been observed in the first patients after 2 years. A low level of local mini-dystrophin expression has been observed in 2 of the 6 patients with some immune response observed against epitopes that are originally absent in the patients [8]. Locoregional HLV administration of AAV is also envisaged. Studies in the rodent, the GRMD dog and nonhuman primates are in progress as well as a clinical study of HLV administration of saline buffer in BMD (Becker Muscular Dystrophy, a mild form of DMD) volunteers. Other applications targeting limb-girdle muscular dystrophy, (alpha or gamma sarcoglycanopathies in particular) are already envisaged (phase I trials of local injections have recently been accomplished) [9, Herson et al., submitted].

Several obstacles remain: technical (large-scale vector production, currently being resolved) or scientific, in particular control of antivector neutralization in subjects who are seropositive for AAV, or an anti-transgene immune response (which occurs in some patients). New vectors with the potential of systemic biodistribution are now emerging. This includes scAAV9 that has been shown to transduce both muscle and motoneurons following systemic administration [10] which opens the way for applications such as spinal muscular atrophy [11].

DNA Surgery
The risks of random integration of the vectors have led to the evaluation of a new research approach which could be called targeted DNA surgery. Meganucleases

are currently being tested for DMD and other myopathies. Meganucleases are derived from natural endonucleases which create a double-stranded cut in the target sequence on the chromosome. Using a DNA template for the sequence of interest, it is thus possible to induce repair by the nuclear machinery through homologous recombination. One can thus address therapeutic approaches for all sorts of human genetic mutations [12], some of which have been validated in vitro [13, 14]. Several myopathies are currently being considered. However, a rapid overview of the possibility of applications of in vitro corrections of cells in the perspective of targeted cell therapy leads to the conclusion that the in vivo applications will encounter the same obstacles as those found in gene therapy.

Rather than acting at the scale of DNA, one alternative consists in acting downstream, at the level of the gene product. Mutations in genes usually lead to formation of a stop codon, producing severe forms linked to 'null mutations' which destroy the reading frame and prevent the production of any dystrophin at all [15, 16]. The more moderate forms result from mutations which respect the reading frame and are compatible with the production of a shorter dystrophin molecule which may conserve some function [1]. This concept has been approached experimentally in the mouse in which it has been possible to study the impact on muscle function of a whole series of deletions of increasing size [17, 18]. These studies pointed to the fact that not every part of the dystrophin protein is necessary for its function, leading to a therapeutic approach by modification of RNA splicing [19].

RNA Surgery
It has thus been envisaged to modify the pre-messenger RNA splicing by exon skipping in order to restore the reading frame, a natural phenomenon which has been deciphered in the so-called 'revertant' fibers of DMD patients [20]. In this situation, antisense nucleotides complementary to the key splicing sequences are chemically modified to make them insensitive to RNases which permit the production of a truncated dystrophin in patients in which there is a complete deficit in dystrophin, i.e. to transform a DMD-type dystrophy into a BMD-type disease. This method of 'splicing therapy' should in theory correct a large number of cases of frameshift or nonsense mutations, by the judicious choice of the exon or the exons to eliminate. The persistence of the therapeutic effect is obtained by repeated administration of the antisense oligonucleotides (their half-life in the cell is only a few weeks) or by vectorization of sequences which produce these antisense oligonucleotides in an AAV or a lentivirus [21]. The first attempt to do this in humans consisted in the intravenous administration of 0.5 mg/kg weekly for 4 weeks of an antisense phosphorothioate destined to skip exon 19 in a 10-year-old DMD patient with a deletion of exon 20. Although the level was low, the presence of in-phase mRNA without exons 19 and 20 in blood lymphocytes and of traces of dystrophin in a muscle biopsy was observed 1 week after the last injection [22].

Two other molecules targeting exon 51 are now in phase IIb trials: (1) PRO051 developed by the Prosensa company has shown dose-dependent expression of quasi-dystrophin in muscle biopsies in two trials [23, 24]. No anti-dystrophin immune response was detected. Long-term follow-up of these patients is in progress. A phase IIb and a phase III trial targeting exon 51 and a phase I/II trial targeting exon 44 are in progress.

(2) AVI BioPharma is developing a morpholino which is reputed to be more stable chemically than the phosphorothioates, AVI-4658. After good results in the GRMD dog, data from clinical trials of intramuscular and then intravenous injections show dose-dependent high levels of expression of quasi-dystrophin in ambulatory patients aged 5–15 years [25].

Exon-skipping must be adapted to the precise genetic characteristics of each patient. Another limitation is represented by the relative incapacity of these products to target the heart, except perhaps AAV-U7 or U1 if adequate conditions for administration become available, and new chemistries developed [26]. A large proportion of patients are not eligible for exon-skipping because some parts of the protein cannot be deleted [1, 27]. Furthermore, since each antisense nucleotide must be considered as a new medication in itself, it will be indispensable to adapt the regulatory constraints to facilitate their development and their accessibility to rare mutations [28]. Alternatively, trans-splicing and skipping of multiple exons are being explored and antisense strategies destined to interfere with the expression of RNA are also being developed to interfere with the expression of RNA in other neuromuscular diseases (dystrophinopathies, myotonic dystrophy and, more generally, all pathologies characterized by triplet repeats for which the accumulation of 'abnormal' RNA has been shown to be toxic) [29].

Forced Readthrough of Stop Codons. Another strategy of splicing therapy involves bypassing premature stop codons induced by a nonsense mutation (about 10% of DMD cases). The principle, which was discovered as an unexpected effect of aminoglycoside antibiotics, was developed using non-antibiotic molecules such as PTC 124® (Ataluren), which is currently in a phase III trial for DMD and cystic fibrosis [30]. The problem here is the specificity of action of the product which must respect not only the natural stops which signal the end of the protein, but also the very numerous premature stops caused by random alternative splicing. A long-term follow-up of patients is indispensable for this strategy. Ten percent of genetic disorders are eligible for this type of approach.

It is possible that for patients in which the disease has already progressed to a late stage that these approaches will be insufficient, mainly because of the extensive muscle degeneration and replacement by fibrosis and adipose tissue (as soon as the regenerative capacities of the muscle are exceeded or exhausted). A regenerative strategy may make it possible to get around this problem.

Regenerative Therapies

Stem Cells

The principle consists of delivering a sufficient number of stem cells with myogenic potential to the muscles. The first clinical trials based on transplantation by local injections of myoblasts, the adult muscle stem cells, were disappointing. Myoblasts contribute to muscle repair in normal and pathologic muscle. Transplantation of healthy myoblasts would make it possible to increase dystrophin production in muscle transplanted into mdx mice. However, multiple clinical trials carried out at the time did not demonstrate the expected gain of function, notably because of rejection of the grafted cells or dystrophin they produced, the limited number of cell divisions of the purified cells, high cell mortality after injection and absence of diffusion of the cells into the injected muscles [31]. Numerous types of cells which might resolve part of the problem have been identified, derived in particular from multiple regions of the organism and often from the vascular zone in which the cell type could be characterized by a collection of membrane markers: satellite cells, muscle-derived stem cells, side population, stem cells derived from bone marrow, mesangioblasts, CD133+ stem cells derived from blood or from muscle and pericytes [32]. Mesangioblasts and CD133+ cells administered into the blood circulation have the capacity to migrate from the vessels and transdifferentiate into skeletal muscle and vascular cells in DMD subjects. The circulating CD133+ cells may even constitute a biomarker for a moderate form of DMD [33]. These cells may be removed from the patients themselves and can be genetically modified as has been shown by exon skipping with lentiviral constructs [34], thus limiting the immunologic risk.

Pharmacologic Induction of Muscle Regeneration. The use of myotrophic growth factors (e.g. IGF-1, growth hormone) has been suggested. More recently, encouraging data have been obtained by targeting the myostatin pathway. Myostatin is a negative regulator of muscle growth. Spontaneous mutations in the myostatin gene produce a hypermuscular phenotype in various bovine and ovine species, as does the inhibition of the gene in the mouse [35, 36]. A mutation which produces the same muscular hypertrophy and unusual force in humans has been found in a German family [37]. Thus, a monoclonal anti-myostatin antibody has been evaluated in a phase I/II trial in 3 neuromuscular disorders including BMD, and a phase II trial has started in DMD patients, using a fusion protein which is an antagonist of the bond between myostatin and its receptor, activin II. Other candidate inhibitors are at the stage of animal studies: gene transfer using the propeptide MPRO in the GRMD dog [38], follistatin [39], or growth and differentiation factor-associated serum protein-1 (GASP-1) [40]. The anti-myostatin strategy could be combined with cytotherapy. Thus the anti-myostatins could favor myogenic differentiation of mesenchymatous stem cells [41].

This strategy could be used alone or preferentially in combination with a therapy which corrects the genetic defect.

Cytoprotection and/or regeneration of motor neurons by embryonic cells or iPS induced to differentiate is the subject of intense research, but is still at a very preliminary stage. Cellular replacement depends on the capacity of grafted cells (of embryonic or iPS origin) to integrate and form new functional connections with their host cells. In the case of spinal muscular atrophy, the motor neurons thus grafted should be capable of forming axonal extensions from the spinal cord to the motor end plate (up to 1 m in humans!). This is possible at the scale of the small animal but does not seem very probable in primates. Research is being oriented instead towards a strategy of neurotrophic support supplied by the cells themselves or after genetic modification [42].

Cytoprotective Strategies

Numerous strategies are being studied which target proteins that are secondarily implicated in the physiopathologic cascade of the genetic deficiency. Proteins which could be modified (by inhibition or stimulation) include utrophin, a protein which is structurally similar to dystrophin. Overexpression of utrophin in patients with Duchenne muscular dystrophy is a natural phenomenon, perhaps compensatory, and one which several groups are trying to amplify pharmacologically (or by gene therapy) to obtain a protective effect on muscle cells [43].

Other targets include proteins linked to the dystrophin complex; other proteins of the muscle membrane, notably alpha 7 integrin; inhibition of proteases, stress signals, or proteins which act on calcium transport or antifibrotics [44]. New pharmacologic targets which come from a more and more precise knowledge of molecular mechanisms are approaching the clinical stage. This is in particular the case of inducers of SMN2, a semifunctional protein which may be able to compensate for the absence of SMN1 which is responsible for infantile spinal muscular atrophy [45]. These pharmacological molecules may selectively stimulate the *smn2* gene, or stabilize its messenger or its protein product.

Pharmacotherapy thus represents an important lead to follow. The only approved treatments for DMD/BMD are (1) glucocorticoids for which the delaying effect on muscle destruction seems to be recognized in both ambulatory and nonambulatory patients [46] although the mechanisms are not yet completely elucidated (anti-inflammatory, induction of muscle regeneration, induction of utrophin, protease inhibition), and (2) cardioprotectors (inhibition of conversion enzyme, β-blockers and more recently idebenone [47, 48], which leads to a considerable reduction in the incidence and severity of the cardiac insufficiency).

Analysis of the pathologic transcriptome may lead to new therapeutic strategies. Several dozens of molecules and nutritional supplements have already been proposed and this leads to the question of the number of clinical trials and patients necessary for their validation.

Conclusions

Neuromuscular diseases are especially complex to approach/treat from a therapeutic point of view. However, they are the target of numerous approaches which are particularly innovative: gene therapy by transfer of complete or partial coding sequences of therapeutic genes into muscular or nervous cell targets, 'surgery' of RNA (splicing therapy or readthrough of stop codons) or DNA, transfer or induction of other genes with a therapeutic perspective. In parallel, the stimulation of tissue regeneration implies embryonic or stem cells which, when induced to differentiate and when administered into the blood circulation, have the capacity to migrate out of the blood vessels and can transdifferentiate. They can also be obtained by pharmacological induction of endogenous cells. Various pharmacological approaches aim at protecting the neuronal, muscular or cardiac cells from degradation. Future treatments will probably involve a combination of specific genetic therapy with nonspecific regenerative and/or cytoprotective therapy.

References

1　Aartsma-Rus A, Van Deutekom JC, Fokkema IF, Van Ommen GJ, Den Dunnen JT: Entries in the Leiden Duchenne muscular dystrophy mutation database: an overview of mutation types and paradoxical cases that confirm the reading-frame rule. Muscle Nerve 2006;34:135–144.

2　Braun S: Muscular gene transfer using nonviral vectors. Curr Gene Ther 2008;8:391–405.

3　Romero NB, Braun S, Benveniste O, et al: Phase I study of dystrophin plasmid-based gene therapy in Duchenne/Becker muscular dystrophy. Hum Gene Ther 2004;15:1065–1076.

4　Hagstrom JE: Plasmid-based gene delivery to target tissues in vivo: the intravascular approach. Curr Opin Mol Ther 2003;5:338–344.

5　Vigen KK, Hegge JO, Zhang G, Mukherjee R, Braun S, Grist TM, Wolff JA: Magnetic resonance imaging-monitored plasmid DNA delivery in primate limb muscle. Hum Gene Ther 2007;18:257–268.

6　Duan D: Myodys, a full-length dystrophin plasmid vector for Duchenne and Becker muscular dystrophy gene therapy. Curr Opin Mol Ther 2008;10:86–94.

7　Wang B, Li J, Xiao X: Adeno-associated virus vector carrying human minidystrophin genes effectively ameliorates muscular dystrophy in mdx mouse model. Proc Natl Acad Sci USA 2000;97:13714–13719.

8　Mendell JR, Campbell K, Rodino-Klapac L, Sahenk Z, Shilling C, Lewis S, Bowles D, Gray S, Li C, Galloway G, Malik V, Coley B, Clark KR, Li J, Xiao X, Samulski J, McPhee SW, Samulski RJ, Walker CM: Dystrophin immunity in Duchenne's muscular dystrophy. N Engl J Med 2010;63:1429–1437.

9 Mendell JR, Rodino-Klapac LR, Rosales-Quintero X, Kota J, Coley BD, Galloway G, Craenen JM, Lewis S, Malik V, Shilling C, Byrne BJ, Conlon T, Campbell KJ, Bremer WG, Viollet L, Walker CM, Sahenk Z, Clark KR: Limb-girdle muscular dystrophy type 2D gene therapy restores alpha-sarcoglycan and associated proteins. Ann Neurol 2009; 66:290–297.

10 Duque S, Joussemet B, Riviere C, Marais T, Dubreil L, Douar AM, Fyfe J, Moullier P, Colle MA, Barkats M: Intravenous administration of self-complementary AAV9 enables transgene delivery to adult motor neurons. Mol Ther 2009;17:1187–1196.

11 Dominguez E, Marais T, Chatauret N, Benkhelifa-Ziyyat S, Duque S, Ravassard P, Carcenac R, Astord S, de Moura AP, Voit T, Barkats M: Intravenous scAAV9 delivery of a codon-optimized SMN1 sequence rescues SMA mice. Hum Mol Genet 2010;Dec 1.

12 Epinat JC, Arnould S, Chames P, Rochaix P, Desfontaines D, Puzin C, Patin A, Zanghellini A, Pâques F, Lacroix E: A novel engineered meganuclease induces homologous recombination in yeast and mammalian cells. Nucleic Acids Res 2003;31:2952–2962.

13 Arnould S, Perez C, Cabaniols JP, Smith J, Gouble A, Grizot S, Epinat JC, Duclert A, Duchateau P, Pâques F: Engineered I-CreI derivatives cleaving sequences from the human XPC gene can induce highly efficient gene correction in mammalian cells. J Mol Biol 2007;371:49–65.

14 Redondo P, Prieto J, Muñoz IG, Alibés A, Stricher F, Serrano L, Cabaniols JP, Daboussi F, Arnould S, Perez C, Duchateau P, Pâques F, Blanco FJ, Montoya G: Molecular basis of xeroderma pigmentosum group C DNA recognition by engineered meganucleases. Nature 2008;456:107–111.

15 Den Dunnen JT, Grootscholten PM, Bakker E, Blonden LA, Ginjaar HB, Wapenaar MC, van Paassen HM, van Broeckhoven C, Pearson PL, van Ommen GJ: Topography of the Duchenne muscular dystrophy (DMD) gene: FIGE and cDNA analysis of 194 cases reveals 115 deletions and 13 duplications. Am J Hum Genet 1989;45:835–847.

16 Koenig M, Hoffman EP, Bertelson CJ, Monaco AP, Feener C, Kunkel LM: Complete cloning of the Duchenne muscular dystrophy (DMD) cDNA and preliminary genomic organization of the DMD gene in normal and affected individuals. Cell 1987;50:509–517.

17 Harper SQ, Hauser MA, DelloRusso C, et al: Modular flexibility of dystrophin: implications for gene therapy of Duchenne muscular dystrophy. Nat Med 2002;8:253–261.

18 Wells DJ, Wells KE, Asante EA, Turner G, Sunada Y, Campbell KP, Walsh FS, Dickson G: Expression of human full-length and minidystrophin in transgenic mdx mice: implications for gene therapy of Duchenne muscular dystrophy. Hum Mol Genet 1995;4:1245–1250.

19 Hoffman EP, Bronson A, Levin AA, Takeda S, Yokota T, Baudy AR, Connor EM: Restoring dystrophin expression in Duchenne muscular dystrophy muscle progress in exon skipping and stop codon read through. Am J Pathol 2011;179:12–22.

20 Matsuo M, Masumura T, Nishio H, Nakajima T, Kitoh Y, Takumi T, Koga J, Nakamura H: Exon skipping during splicing of dystrophin mRNA precursor due to an intraexon deletion in the dystrophin gene of Duchenne muscular dystrophy kobe. J Clin Invest 1991; 87:2127–2131.

21 Goyenvalle A, Vulin A, Fougerousse F, Leturcq F, Kaplan JC, Garcia L, Danos O: Rescue of dystrophic muscle through U7 snRNA-mediated exon skipping. Science 2004;306:1796–1799.

22 Takeshima Y, Yagi M, Wada H, Ishibashi K, Nishiyama A, Kakumoto M, Sakaeda T, Saura R, Okumura K, Matsuo M: Intravenous infusion of an antisense oligonucleotide results in exon skipping in muscle dystrophin mRNA of Duchenne muscular dystrophy. Pediatr Res 2006;59:690–694.

23 van Deutekom JC, Janson AA, Ginjaar IB, et al: Local dystrophin restoration with antisense oligonucleotide PRO051. N Engl J Med 2007;357:2677–2686.

Piguet P, Poindron P (eds): Genetically Modified Organisms and Genetic Engineering in Research and Therapy. BioValley Monogr. Basel, Karger, 2012, vol 3, pp 86–102

Gene Therapy for Cancer Treatment – State of the Art

Amor Hajri

Institute of Biology III, University of Freiburg, Freiburg, Germany

Abstract

During the last decades, significant advances have been made in the field of gene therapy for the identification of new therapeutic genes, the design of innovative vectors and different ways of targeting. Cancer gene therapy is not indicated in clinical practice yet. However, basic and clinical advances have been reported which indicate the swift evolution of this field which, without any doubt, will be part of future cancer therapies. Many of the past obstacles and barriers are being actively overcome now. We are hopeful that cancer gene therapy will continue to progress and ultimately take its place within the existing anticancer armamentarium.

Advances in molecular medicine as well as completion of the human genome project have allowed the identification of numerous disease-causing genes. These led to a new therapeutic approach, i.e. gene therapy that uses genes as a medicine to treat, cure or even prevent disease and thereby improve the quality of life.

Gene therapy includes the treatment of both genetically based and infectious diseases by introducing genetic materials which have therapeutic effects [1–3]. It was first conceived as a treatment for hereditary single-gene defects, but gene replacement or repair can bring permanent relief in a number of other disorders. Therefore, gene therapy is also being developed for (1) acquired diseases such as cancer, (2) cardiovascular diseases, or (3) neurodegenerative disorders including Alzheimer's, Parkinson's and Huntington's diseases, and (4) infectious diseases. In the future, gene therapy approach may allow doctors to treat a disorder by inserting a gene into a patient's cells instead of using drugs or surgery.

Although gene therapy is a promising treatment strategy for a number of diseases, it is still considered an experimental strategy and the technique remains risky and is still under study to assess its safety and effectiveness. Gene therapy is currently only being tested for the treatment of diseases at an advanced stage that have no other cure.

What Is Gene Therapy?

The science of gene therapy relies on the introduction of genes to cure a defect or slow down the progression of a disease and thereby improve the quality of life. Today, gene therapy is still considered an experimental strategy. But, its extraordinary evolution suggests that in the near future, this approach may allow doctors to treat a disorder by inserting a gene into patient's cells instead of using conventional drugs or surgery.

Researchers actively study and test several options for gene therapy applications, including:
- Replacing a mutated gene that causes disease with a healthy copy of the corresponding gene.
- Inactivating, or 'knocking out', a mutated gene that is functioning improperly.
- Introducing a new gene into the body to help fight a disease.

During the last decades, intensive research in the area of gene therapy has been conducted worldwide with the first approved gene therapy clinical trial in 1990. In this study, the adenosine deaminase (ADA) gene was transferred into T cells of two children with severe combined immunodeficiency (ADA-SCID) [4].

Several hundred clinical studies in gene therapy have now been performed. These trials have shown that there are still several major factors that prevent gene therapy from becoming a routine way to treat genetic conditions and disorders. Cancer gene therapy represents one of the most rapidly developing areas in pre-clinical and clinical cancer research. However, some problems need to be solved before this strategy becomes routinely adopted in clinic. Two of the most important problems are low efficiency of gene transfer and lack of selectivity of the existing vectors.

Gene Transfer and Vectors

Even though gene therapy holds great promise for the achievement of this task, the transfer of genetic material into higher organisms remains an enormous technical challenge. The most important question to answer is how to deliver and target gene expression specifically, efficiently and safely? During the last years, several delivery systems have been developed and some of them are currently being tested in clinical trials.

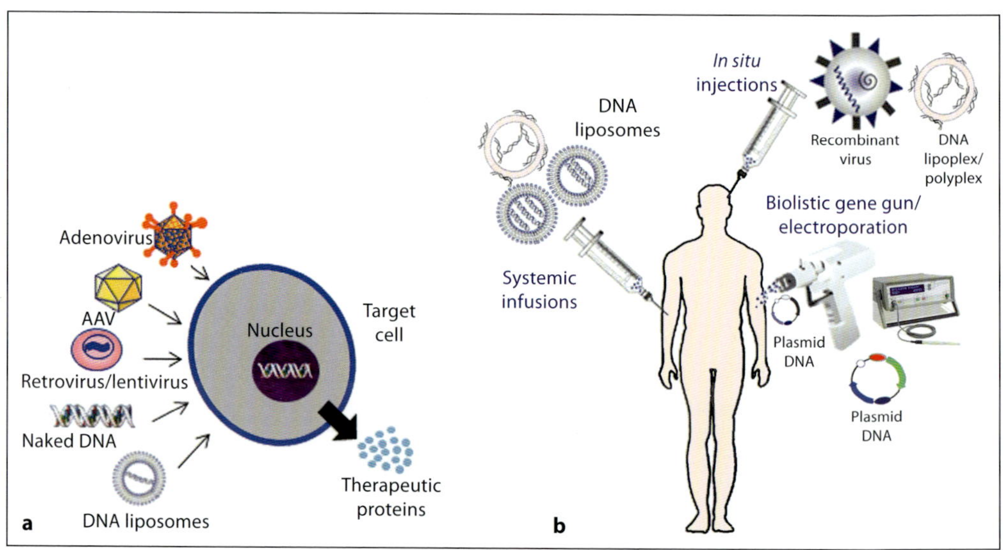

Fig. 1. Different vectors in use for gene therapy. Viral and nonviral vectors for gene delivery in target cells to produce the therapeutic proteins (**a**), and in vivo vector delivery using different strategies (topical and systemic applications) (**b**).

The presently available gene delivery methods utilize viral and non-viral (or synthetic) vectors, as well as physical approaches [5–7] (fig. 1). Viruses offer greater efficiency of gene delivery; however, nonviral vectors are preferred due to safety concerns with the viral vectors [6–9]. For in vivo applications, nonviral gene therapy is preferred over viral therapies for being less or nonimmunogenic and nonintegrating into the host genome, despite a less sustained gene expression and low amounts of protein production, as shown in most studies. During the last years, considerable effort has focused on nonviral gene expression systems for improving not only gene carriers but also plasmid design. This includes removal of inessential bacterial CG dinucleotides (CpGs), the incorporation of scaffold/matrix attached region (S/MAR) sequences to prolong expression, promoter selection for specific gene expression, and improving plasmid entry into the nucleus [10]. Compared with nonviral systems, viral vectors are much more efficient in delivering exogenous genes to target cells and inducing long-term gene expression.

Viral vector-mediated gene transfer employs replication-competent and -deficient viruses. A major advantage of viral vectors is their high gene delivery efficiency. A variety of virus vectors have been employed to deliver genes to cells to provide either transient (e.g. adenovirus, vaccinia virus) or permanent (e.g. retrovirus, adeno-associated virus) transgene expression and each approach has its own advantages and disadvantages (table 1). However, not all viruses are

Table 1. Some advantages and disadvantages of different gene therapy vectors

Vector	Advantages	Disadvantages
Viral vectors		
Adenovirus	Large transgene insert capacity Biologically safe	High immunogenicity
Herpes simplex virus	Large transgene insert capacity Availability of anti-herpetic drugs	Immunogenicity
Adeno-associated viruses	Stable transgene expression Reduced immunogenicity	Small transgene insert size capacity Need for helper viruses during manufacture
Retrovirus	Stable transgene expression Only infects dividing cells	Small transgene insert size capacity Possible insertional mutagenesis (integration)
Lentivirus	Infects both dividing cells and non-dividing cells	Possible insertional mutagenesis (integration)
Vaccinia virus	Long history of safe human use Large transgene insert capacity	High immunogenicity Productive infection in immune suppressed patients Replication in skin lesions (eczema, psoriasis)
Vesicular stomatitis virus	Selective replication-competence in cells with defective interferon response (tumor cells)	High immunogenicity Animal pathogen (safety/environmental concerns)
Measles virus	Long history of safe human use (Edmonston vaccine strain)	Most adults are immune Wild-type virus is immunosuppressive Rare measles-like illness with vaccine strains
Non-viral vectors		
Naked DNA	"Easy" engineering Low immunogenicity	Rapid clearance, low transfection efficiency in vitro
Synthetic vectors	Large gene carrying capacity Nanoparticles can accumulate into tumors as a result of the enhanced permeability and retention effect	Cationic liposomes have an inflammatory toxicity and a low transfection efficiency in vivo

Adapted from Touchefeu Y, et al: Aliment Pharmacol Ther 2010;32:953–968.

suitable for gene delivery. Factors for consideration for the choice of a particular vector include the packing capacity of transgenes, host range, tropism of vectors, inflammatory potential of the viral vectors, and serotypes [11, 12]. Another very important advantage is that some viruses are not only used as shuttles but as oncolytic viruses (more details in the section on virotherapy).

The nonviral vectors developed so far principally include cationic liposomes, cationic polymers and synthetic peptides, which can complex with negatively charged nucleic acids to form particles with a diameter in the order of 100 nm. The nonviral vectors appear to be highly effective in gene delivery in vitro in cell cultures but are significantly less effective in vivo. Physical methods utilize mechanical pressure, electric shock or hydrodynamic force to transiently permeate the cell membrane to transfer DNA into target cells. They are simpler than viral- and nonviral-based systems and highly effective for localized gene delivery. The past decade has seen significant efforts to establish the most desirable method for safe, effective and target-specific gene delivery, and good progress has been made.

Cancer Gene Therapy

This review provides insights into the state-of-the-art accomplishments made with gene-based therapies. The review will focus on cancer gene therapy applications describing the general approaches and providing a summary of the recent advances and will discuss some examples of gene delivery and targeting strategies.

Despite improved diagnostic methods and therapeutic regimens, cancer has a poor prognosis in the majority of cases. There is an increasing demand for novel diagnostic and therapeutic approaches. From new insights into the molecular basis of tumorigenesis and from the advent of recombinant DNA technology, the opportunity to treat cancer by using genetic information has emerged.

Gene therapy has the potential to provide cancer treatments based on novel mechanisms of action with potentially low toxicities. However, application of this type of cancer treatment requires an understanding of tumor biology, methods for delivering genes to specific cell types, and strategies to regulate the level and duration of gene expression. Several strategies have been developed and ongoing clinical protocols can be divided into four groups: (1) suppression of oncogenes or transfer of tumor-suppressor genes; (2) enhancement of immunological response; (3) transfer of suicide genes, and (4) protection of bone marrow using drug resistance genes.

The therapeutic effect of the gene should be targeted and last as long as required in an appropriately regulated fashion. In recent years, there have been intensive efforts to generate targetable, injectable vectors based on a variety of viral and nonviral gene delivery systems. Then, it is expected that this therapy may provide more effective control of locoregional recurrence in diseases as well as systemic control of micrometastases.

Tumor Targeting
The potential of gene therapy to target the expression of therapeutic genes to the desired target cells makes it particularly attractive for cancer treatment. Some progresses have been made in vector targeting with viral (nonreplicative and replication-competent) and nonviral vectors.

Targeted expression of therapeutic genes is essential to prevent this toxicity. Recent understanding of tissue- and cancer-specific gene regulation in organogenesis and cancer fields has enabled the development of new promoter/enhancer systems to express therapeutic genes only in targeted cells and tissues. Cancer-specific therapeutic gene expression reduces undesirable toxicity.

From the viewpoint of clinical application, a safe and efficient gene therapy applied only to the target tumor tissues is of utmost importance. Although targeting does not seem to be important for intratumoral gene delivery, it becomes crucial when systemic gene transfer is performed. Then tumor-specific gene transfer can improve the therapeutic effect by preventing damage of healthy tissues and decreasing the risk of germ-line transduction. Several methods have been explored to increase the specificity and to target tumor cells.

They can be roughly divided into two major categories. Targeted gene therapy of malignancies can be achieved through targeted gene delivery or targeted gene transcription (expression). In the first approach the recombinant vector is targeted at the level of transduction to achieve the selective delivery of the therapeutic gene in tumor cells [6, 13, 14]. Recent advances in targeted delivery include the successful use of bifunctional crosslinkers to target adenoviral and retroviral vectors, inserting short targeting peptides and larger polypeptide-binding domains into the coat proteins of a number of different viral vectors, and replication-competent vectors which have been shown to be promising as anti-cancer agents. Some other nonviral therapeutic agents, including receptor-mediated DNA or liposome-DNA complex, and bacteria vehicles have also been developed.

In the second alternative approach for the derivation of cancer-cell-specific vectors, therapeutic gene expression is targeted at the level of transcription. This is accomplished by placing the therapeutic gene under the control of tissue or tumor-specific transcriptional regulatory sequences that are activated in tumor cells but not in normal cells and therefore target expression selectively to the tumor cell [15–21]. Tumor-specific promoters or regulatory elements such as carcinoembryonic antigen (CEA), mucin1 (MUC1), alpha-fetoprotein (AFP) or the human telomerase reverse transcriptase (hTERT) promoters have the potential to target toxic proteins or suicide genes to gastrointestinal tumors. MUC1 and CEA are normally expressed in fetal tissue and are transcriptionally silent in adults but are frequently overexpressed in human adenocarcinoma. Human telomerase reverse transcriptase, the catalytic subunit of the telomerase, is virtually undetectable in most normal tissues and its activity is significantly higher in approximately 90% of cancers, and correlates well with the degree of malignancy [16, 17]. By reducing cytotoxicity in normal cells, the use of specific promoters provides a clear advantage over the constitutive promoters such as the human cytomegalovirus (CMV) and the phosphoglycerate kinase (PGK) promoters.

Molecular Approaches in Cancer Gene Therapy
Upregulation or downregulation of some genes is the basis of tumor initiation and progression. The underlying mechanism of gene dysfunction includes many mutations on the genetic level resulting in molecular and cellular changes that are specific of cancer development. Many genes are involved in disruption of balanced harmony of cellular networks and gene expression programs that maintain the cellular homeostasis. There are principally two major groups, including oncogenes and tumor suppressor genes. Thus, it is not surprising that the majority of cancer gene therapy approaches involve strategies either to suppress the function of activated oncogenes or to restore the expression of functional tumor suppressor genes. Other approaches are intensively developed to potentiate the antitumor activity of the immune system, to downregulate angiogenesis and metastasis, or to initiate tumor self-destruction [22, 23].

One of the most promising approaches emerging from the improved understanding of cancer at the molecular level is the possibility to use gene therapy to selectively target and destroy tumor cells, for example, by targeting the loss of tumor suppressor genes (e.g. the P53 gene) and the overexpression of oncogenes (e.g. K-RAS) that have been identified in several malignancies. Here we discuss some data using p53 and KRAS-based cancer gene therapy.

P53 Tumor Suppressor Gene Therapy for Cancer
During the last decade, tumor suppressor genes have been widely used in cancer treatments. Examples of tumor suppressor genes involved in this process include p53 (regulates cell cycle and apoptosis), retinoblastoma gene Rb (regulates cell cycle and differentiation), p16INK/CDKN2 (regulates cell cycle), and PTEN (regulates cell survival). By inhibiting the expression of tumor suppressor genes, cancer cells with DNA mutations can continue to proliferate, acquire new gene mutations, become less differentiated, and escape apoptosis.

Currently, it is well established that the p53 tumor suppressor gene plays a critical role in safeguarding the integrity of the genome and preventing tumorigenesis. Although multiple genes are involved in carcinogenesis, mutations or deletions of the p53 (TP53) gene are the most frequent abnormality identified in human tumors. Preclinical studies both in vitro and in vivo have shown that restoring p53 function can induce apoptosis in cancer cells. The *TP53* gene is one of the most studied genes in human cancer [23, 24]. Its normal role is to halt the division of a defective cell and then force the cell to kill itself. In recent years, considerable interest was focused on mutant p53, the abnormal protein product of *p53* mutations that often accumulate in cancer cells. There is now compelling experimental evidence that many mutations can exert mutant-specific, gain-of-function effects by perturbing the regulation of expression of multiple genes [24, 25].

Clinically, *p53* mutations have been associated with poor prognosis and drug resistance in a growing array of malignancies. It is expected that reactivation of the wild-type conformation of p53 should induce massive apoptosis in tumor cells. There have already been many attempts at restoring or modulating p53 protein functions through genetic or pharmacological approaches.

Even if clinical attempts to replace the p53 gene have had limited success, progress is being made in the development of new therapeutic approaches targeting p53 alterations.

The first p53-based gene therapy was reported in 1996. A retroviral vector containing the wild-type p53 gene under the control of an actin promoter was injected directly into tumors of non-small cell lung cancer patients [25].

A few trials reached phase III, but final approval from the FDA has not yet been granted. Although preliminary results were promising, recent data failed to demonstrate antitumor activity in patients and some trials have been discontinued [26]. New trials aim at combining gene transfer with chemotherapy or radiotherapy.

Recently, p53-based gene therapy has been developing in China [27]. There are ongoing investigations into the use of *p53*-based gene therapy in cancer [28, 29], and clinical approval for a replication-defective, *p53*-producing adenovirus, Gendicine (Shenzhen Sibiono Genetech, Shenzhen, China), was obtained in China in 2003. Extensive trials with the equivalent Advexin® failed to reach their efficacy end points despite some clear responses in some patients [30, 31].

KRAS Antisense Oligonucleotides and Small Interfering RNA
Expression of proto-oncogenes during fetal development is important for generating enough cells to form the various organs in the body. Most proto-oncogenes are then silenced to prevent abnormal tissue growth. Often cancer cells propagate by activating and amplifying proto-oncogenes (oncogenes).

One form of cancer gene therapy is the targeted disruption of tumor oncogenes.

For oncogene overexpression, specific antisense oligonucleotides (aODN) which bind and subsequently inhibit oncogene activity (i.e. antioncogenes) can also be utilized in cancer therapy. More recently, small interfering RNA (siRNA) studies for post-transcriptional oncogene silencing showed considerable interest in cancer therapy [32–34].

Thus, the overexpression of an oncogene such as KRAS, one of the most studied proto-oncogenes, can be blocked at the genetic level by integration of an antisense sequence whose transcript binds specifically to the oncogene

RNA, disabling its capacity to produce protein. In vitro and in vivo experiments have demonstrated that when an antisense KRAS vector is integrated into cancer cells that overexpress KRAS their tumorigenicity is decreased [35–37]. In our laboratory, using experimental pancreatic tumors, we demonstrated that specific aODN targeting the mutated KRAS inhibited endogenous KRAS expression and reduced tumor cell proliferation by 30–40%. These effects were restricted to the relevant KRAS mutated cells [37]. Additionally, we demonstrated that specific siRNAs targeting mutated KRAS induced better effects than aODNs. In vivo studies have shown that treatment with KRAS-specific aODN and siRNA induce significant inhibition of tumor growth resulting in a significant improvement in the survival of mice bearing orthotopic pancreatic xenografts, without any apparent signs of toxicity [37]. Moreover, combination of shRNA plasmids expressing KRAS siRNA with gemcitabine resulted in a remarkable increase in antitumor response and survival prolongation of mice receiving both therapies compared with those treated with gemcitabine alone. These findings have strong implications for pancreatic cancer therapeutic strategies.

Despite some disappointments with the technology, advances in the medicinal chemistry and formulation of antisense oligonucleotides and siRNAs will further enhance their therapeutic potential in the future. Moreover, for long-term expression and targeting, several researchers have designed recombinant vectors expressing short hairpin RNA (shRNA) [38]. Vectors expressing shRNA are attractive for efficient and tissue-specific RNA interference delivery. The big advantage of siRNA approach in oncology is that it can theoretically target any gene which is related to cancer development and progression (including tumor cell invasion, metastasis and tumor angiogenesis).

Apoptosis and Cancer Gene Therapy
In addition to the essential role of apoptosis in the homeostasis of normal tissue, there is increasing evidence that the processes of neoplastic transformation, progression and metastasis involve alterations in the normal apoptotic pathways. Furthermore, the majority of chemotherapeutic agents as well as radiation utilize the apoptotic pathways to induce cancer cell death. It was demonstrated that several types of cancers are characterized by defects in apoptosis or in the apoptotic regulatory pathways such as p53, the nuclear factor kappa B (NFκB), or phosphatidylinositol 3-kinase (PI3K)/Akt leading to defects in apoptosis [39]. Moreover, the failure of chemotherapy can be attributed, at the cellular level, to an acquired resistance to apoptosis. Because these pathways may be preferentially altered in tumor cells, there is potential for a selective effect in tumors, thus sparing normal tissue. Therefore, devising mechanisms to restore the apoptotic process in tumor cells may help in the design of effective therapeutic strategies against resistant cancers [40–42]. Several therapeutic agents have been developed which target the apoptotic machinery and some of

them are in clinical use or in clinical trials, while others are still in preclinical and experimental stages [43–45]. Generally, these agents are designed to target the extrinsic pathway, the intrinsic pathway, the common pathway (through caspases), or the proteins regulating apoptosis. In cancer gene therapy, the activation of these apoptotic pathways can be induced through the delivery of specific expression vectors. It is important that these proapoptotic genes are controlled and targeted in tumor cells. Several studies were performed and reported in the literature indicate the potential interest of targeting apoptosis pathways in cancer therapy. In our laboratory, we designed inducible recombinant adenoviruses expressing Bax or TRAIL under the control of hTERT promoter. In these studies with pancreatic adenocarcinoma models, our data showed that treatment with these adenoviral systems resulted in high-level expression of *Bax* and *TRAIL* genes directly related to apoptosis induction, leading to a significant sensitization of chemoresistant pancreatic tumor cells. Furthermore, treatment with Bax/TRAIL adenoviruses plus a suboptimal dose of gemcitabine resulted in significant tumor regression and prolongation of the experimental animal's life, in contrast to the weak retardation in tumor growth observed when gemcitabine alone was used [41]. These data and others reported by authors worldwide suggest that therapy regimens using targeted suicide-apoptogene therapy will provide new promise for cancer treatment in the future.

Suicide Gene Cancer Therapy
Gene-directed enzyme/prodrug therapy (GDEPT) or suicide gene therapy is an attractive alternative approach to cancer therapy, with the potential to give therapeutic ratios superior to standard chemo- and radiotherapy. The suicide genes encode enzymes which are not toxic per se, but which catalyze the formation of highly toxic metabolites following the application of a much less toxic prodrug [46] (fig. 2).

The transfer of drug-activating enzyme gene into tumor cells and treatment with a prodrug form of chemotherapeutic agents causes a high concentration of the activated toxic drug in the tumor tissue and apoptosis of tumor cells. Furthermore, the importance of this approach is related to the fact that not only transduced cells, but also circumferential or neighboring cells are reported to die with this gene therapy (bystander effect) [47]. For most of the solid tumors, the transfection efficiency represents an obstacle for gene therapy to eradicate tumor cell growth. Since expression of foreign suicide genes will not occur in all cells of a solid tumor, a bystander effect is required, allowing the produced cytotoxic metabolites to kill not only the transduced tumor cells but also the neighboring untransduced tumor cells. Therefore, an extended bystander effect, apoptosis induction, and immune response would be advantageous characteristics of suicide gene/prodrug system for the treatment of cancer in general and particularly in solid tumors. Several reports indicated the critical role of

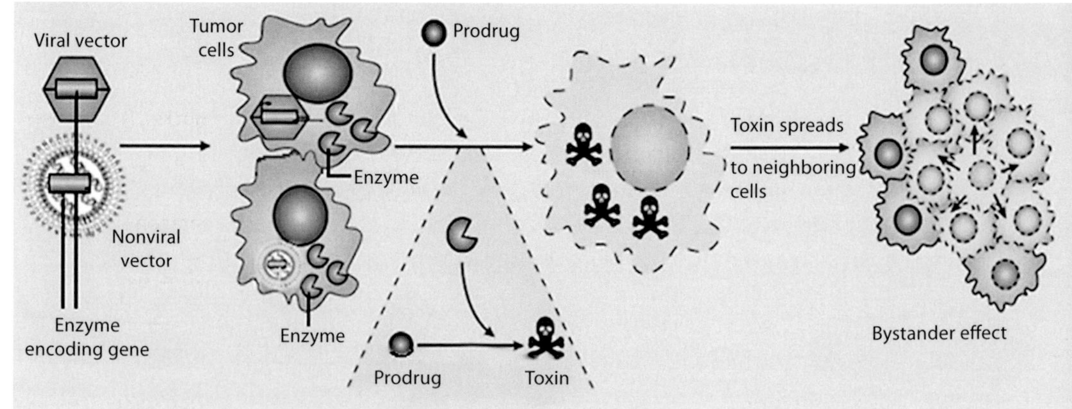

Fig. 2. Mechanism of suicide gene transfer showing principle of gene-directed enzyme prodrug therapy. The vector delivers a suicide gene that encodes a prodrug-converting enzyme to tumor cells. Then delivery of the non-toxic prodrug to the tumor cells results in (1) prodrug conversion to the active, cytotoxic metabolite in the tumor cell, and (2) diffusion to neighbouring cells conferring a potent bystander effect. Adapted from McCormick [3].

'bystander effect' which varied from one system to another depending on the activated drug diffusion and on the condition of the intercellular gap junctions [47, 48].

During the last decade, several enzyme/prodrugs GDEPT systems have been optimized (table 2). Two GDEPT systems that have been investigated extensively are the herpes simplex virus thymidine kinase/ganciclovir (hsv-TK/GCV) and the *E. coli* cytosine deaminase/5-fluorocytosine (eCD/5-FC); both have been tested in clinical trials [46, 49–51]. In our laboratory, we have used these suicide genes for pancreatic cancer and developed another system using *Escherichia coli* purine nucleoside phosphorylase with the methyl purine deoxyriboside (ePNP/MePdR). We demonstrated that ePNP gene expression followed by MePdR treatment induced a cytotoxic effect with a strong bystander effect in pancreatic tumor cell cultures and dramatic antitumor activity in a pancreatic tumor xenograft model [52].This suicide gene/prodrug system has the advantage regarding hsv-TK/GCV and eCD/5-FC, in that it activates prodrugs that are able to kill quiescent as well as proliferating cells [52, 53]. We demonstrated that ePNP-transduced pancreatic tumor cells were sensitive to MePdR treatment, indifferently at 50% confluence (actively proliferating) or 100% confluence (low or no dividing cells), of transduced cells or mixed transduced with untransduced cells (1:1) [52]. Furthermore, in addition to the high cytotoxic activity, this ePNP/MePdR system induced a potent bystander effect on pancreatic tumors. The use of 2–5% of transduced cells can completely abolish tumor growth in cell cultures. In contrast to hsv-TK, the bystander mechanism of

Hajri

Table 2. Selected examples of Enzyme-prodrug combinations for suicide gene therapy

Enzymes	Origins	Prodrugs	Products	Mechanisms
HSV-thymidine kinase	Herpes simplex virus	Ganciclovir	Ganciclovir triphosphate	Blocks DNA synthesis
Cytosine deaminase	*E. coli*/Yeast	5-Fluorocytosine	5-Fluorouracil (5-FU)	Pyrimidine antagonist: blocks DNA and RNA synthesis
Nitroreductase	*E. coli*	CB-1954 and analogs	Hydroxylamine/amine derivatives	DNA crosslinking
Carboxylesterase	Human	CPT-11	SN38	Topoisomerase inhibitor
Cytochrome P450	Human, rat	Cyclophosphamide	Phosphoramide mustard	DNA alkylating agent: blocks DNA synthesis
Purine nucleoside phosphorylase	*E. coli*	6-Mercaptopurine-DR	6-Mercaptopurine	Purine antagonist: blocks DNA synthesis
Deoxycytidine kinase	Human	Cytoarabinoside(Cyto) Gemcitabine(Gem)	Cyto- monophosphate Gem- monophosphate	Antimetabolite: inhibits DNA synthesis Antimetabolite: inhibits DNA synthesis
Carboxypeptidase A	Human	MTX-alpha-peptides	MTX	Antimetabolite: inhibits DNA synthesis

MePdR/ePNP seems to be due to free diffusion of MeP or soluble toxic factors across the cell membrane without the necessity of cell contact to achieve killing of neighboring untransduced cells.

Oncolytic Virotherapy
Virotherapy is the therapeutic use of viruses – both naturally occurring and genetically altered – for the selective destruction of cancerous cells. In recent years, oncolytic viruses have been increasingly studied as potential cancer therapeutics [54–56].

Cancer-selective oncolytic viruses replicate preferentially in cancer cells and, as a result, destroy those cells at the end of replication cycles; normal cells are spared and hence toxicity is limited. However, the native viruses are generally not potent enough to eradicate tumors; they are limited by their ability to eliminate tumors by oncolysis. Some oncolytic viruses are naturally occurring, whilst others are genetically engineered to reduce pathogenicity, enhance tumor cell selectivity and encode therapeutic genes. Naturally occurring oncolytic viruses include: Newcastle disease virus (NDV), vesicular stomatitis virus (VSV), myxoma virus, reovirus, Seneca Valley virus, coxsackie A viruses and echoviruses. Engineered oncolytic viruses encompass backbones from adenoviruses, vaccinia viruses, herpes simplex virus (HSV), parvovirus and poliovirus. To improve and increase the intrinsic antitumor activity, oncolytic viruses can be 'armed' to express exogenous therapeutic genes including cytokines or prodrug-activating enzymes or proapoptotic genes or inhibition of angiogenesis. Exploitation of bystander effects represents one option to the technical hurdle of infecting all the cells in a tumor. A second approach uses the ability of viruses to spread from their site of inoculation and infect neighboring cells. In this approach, cells are killed as a consequence of virus infection, as they become factories for producing new infectious virus particles [57, 58].

It is well demonstrated that cancer-selective oncolytic viruses replicate preferentially in cancer cells and as a result, destroy those cells at the end of replication cycles; normal cells are spared and hence toxicity is limited. Of note, oncolytic viruses can kill apoptosis-resistant tumor cells, and hence do not have cross-resistance with existing therapies. Engineered oncolytic viruses have been developed and have various mechanisms-of-actions: (1) inherently tumor-selective virus species (for example, RNA viruses, poxviruses); (2) viral gene-deleted mutants – critical viral gene expendable for growth in tumor cells, but not in normal cells, were deleted (for example, adenovirus dl1520/Onyx-015, herpes simplex virus (HSV) G207); (3) promoter engineered mutants –viral replication was engineered to be dependent on inserted tumor-specific promoters, and as a result, the replication of the virus is restricted to tumor cells that are able to activate the promoters (for example, prostate specific antigen-regulated adenovirus CG7870, telomerase-regulated adenoviruses and HSVs); (4) pseudotyped viruses – normal viral tropism is ablated, and viruses are engineered to attach/bind to specific surface receptors that are expressed exclusively/preferentially on tumor cells (for example, adenovirus Delta-24RGD). The most critical barrier to the widespread use of this cancer treatment is the deactivation of the host immune system, which quickly develops ways to destroy tailored viruses. Numerous experiments have been done to modify the immune response in favor of virus replication and tumor lysis [59, 60].

Currently, oncolytic viruses are being evaluated in pre-clinical and clinical settings.

The virus should infect, replicate in, and destroy human tumor cells, ideally including noncycling cancer cells. When considering a virus species for development as an oncolytic therapy, a number of efficacy, safety and manufacturing issues needs to be assessed. Nonintegrating viruses have potential safety advantages in that unpredicted events caused by genomic integration are avoided. A genetically stable virus is desirable from both safety and manufacturing standpoints.

Thus, it is desired that an oncolytic virus displays a number of attributes such as:

1   Relatively low pathogenicity.
2   Being easily genetically manipulated.
3   Being able to replicate efficiently and specifically in malignant cells.
4   Being able to be delivered systemically.
5   A lack of serious side effects following administration.
6   The virus must be amenable to high-titer production and purification under the Good Manufacturing Practices (GMP) guidelines for clinical studies.

The recent progress of genetic virus modifications for better viral delivery (infection), tumor selectivity and spreading confirm that virotherapy holds great promise as a novel treatment platform for cancer. Modifications of viruses to enhance tropism for tumors with greater specificity for tumor cell transduction and replication will lead to safer profile and fewer adverse events associated with applications of vectors. Also, the advantages of oncolytic virotherapy include the potential lack of crossresistance with standard therapies and their ability to cause tumor destruction by numerous mechanisms. Additionally, it appears from several studies that the future of oncolytic virotherapy lies in combination therapies (versus monotherapy) for cancer eradication. A number of results were reported demonstrating the rational of this strategy combining chemotherapy or radiotherapy with virotherapy.

The acceptance of oncolytic viruses in the clinic has increased, following the publication of clinical trial results, reporting safety and efficacy data. Moreover, China has recently granted approval for the clinical use of the oncolytic virus, adenovirus H101, specifically for the treatment of head and neck cancer.

Conclusions

The growing understanding of genetics and molecular biology of human diseases has led to the design of more rational therapeutic approaches. In this context, gene therapy offers potential opportunities for new treatment options for many incurable diseases.

This review attempted to highlight the progress made in recent years in using gene therapy to treat cancer. It summarizes the most illustrative examples of cancer gene therapy realized in preclinical experiments and clinical trials. Then, it is clear that many exciting innovations emerge indicating that gene therapy is a promising modality for the treatment of malignant tumors for which conventional therapies are often inadequate. The identification of new efficient target genes, design of smart recombinant vectors for gene transfer and tumor targeting, and development of new experimental models allowed cancer gene therapy to progress considerably from animal experiments to preliminary clinical trials. Objectively, we can be optimistic that gene therapy will be of clinical value in the near future.

References

1 Cross D, Burmester JK: Gene therapy for cancer treatment: past, present and future. Clin Med Res 2006;4:218–227.

2 Qasim W, Gaspar HB, Thrashe AJ: Progress and prospects: gene therapy for inherited immunodeficiencies. Gene Ther 2009;16: 1285–1291.

3 McCormick F: Cancer gene therapy: fringe or cutting edge? Nat Rev 2001;1:130–141.

4 Blaese RM, Culver KW, Miller AD, Carter CS, Fleisher I, Clerici M, Shearer G, Chang L, Chiang Y, Tolstoshev P, Greenblatt JJ, Rosenberg SA, Klein H, Berger M, Mullen CA, Ramsey WJ, Muul L, Morgan RA, Anderson WF: T lymphocyte-directed gene therapy for ADA- SCID: initial trial results after 4 years. Science 1995;5235:475–480.

5 Boulaiz H, Marchal JA, Prados J, Melguizo C, Aranega A: Non-viral and viral vectors for gene therapy. Cell Mol Biol 2005;51:3–22.

6 Guo X, Huang L: Recent advances in nonviral vectors for gene delivery. Acc Chem Res 2011;Oct 26 [Epub ahead of print].

7 Waehler R, Russell SJ, Curiel DT: Engineering targeted viral vectors for gene therapy. Nat Rev Genet 2007;8:573–587.

8 Xu L, Anchordoquy T: Drug delivery trends in clinical trials and translational medicine: challenges and opportunities in the delivery of nucleic acid-based therapeutics. J Pharm Sci 2011;100:38–52.

9 Rogers ML, Rush RA: Non-viral gene therapy for neurological diseases, with an emphasis on targeted gene delivery. J Cont Release 2012;157:183–189.

10 Wu TL, Zhou D: Viral delivery for gene therapy against cell movement in cancer. Adv Drug Deliv Rev 2011;63:671–677.

11 Sheridan C: Gene therapy finds its niche. Nat Biotechnol 2011;29:121–128.

12 Wu TL, Ertl HC: Immune barriers to successful gene therapy. Trends Mol Med 2009; 15:32–39.

13 Lavigne MD, Górecki DC: Emerging vectors and targeting methods for nonviral gene therapy. Expert Opin Emerging Drugs 2006;3:541–557

14 Stone D, David A, Bolognani F, Lowenstein PR, Castro MG: Viral vectors for gene delivery and gene therapy within the endocrine system. J Endocrinol 2000;103–118.

15 Sadeghi H, Hitt MM: Transcriptionally targeted adenovirus vectors. Curr Gene Ther 2005;5:411–427.

16 Shay JW, Zou Y, Hiyama E, Wright WE: Telomerase and cancer. Hum Mol Genet 2001;7:677–685.

17 Hiyama E, Hiyama K: Telomerase detection in the diagnosis and prognosis of cancer. Cytotechnology 2004;45:61–74.

18 Doloff JC, Waxman DJ, Jounaidi Y: Human telomerase reverse transcriptase promoter-driven oncolytic adenovirus with E1B-19 kDa and E1B-55 kDa gene deletions. Hum Gene Ther 2008;19:1383–1400.

19 Goto H, Osaki T, Kijima T, Nishino K, Kumagai T, Funakoshi T, Kimura H, Takeda Y, Yoneda T, Tachibana I, Hayashi S: Gene therapy utilizing the Cre/loxP system selectively suppresses tumor growth of disseminated carcinoembryonic antigen-producing cancer cells. Int J Cancer 2001;94:414–419.

20 Doloff JC, Waxman DJ: Dual E1A oncolytic adenovirus: targeting tumor heterogeneity with two independent cancer-specific promoter elements, DF3/MUC1 and hTERT. Cancer Gene Ther 2011;18:153–166.

21 Qiao J, Doubrovin M, Sauter BV, Huang Y, Guo ZS, Balatoni J, Akhurst T, Blasberg RG, Tjuvajev JG, Chen SH, Woo SL: Tumor-specific transcriptional targeting of suicide gene therapy. Gene Ther 2002;9:168–175.

22 Hughes RM: Strategies for cancer gene therapy. J Surg Oncol 2004;85:28–35.

23 McCormick F: Cancer gene therapy: fringe or cutting edge? Nat Rev 2001;1:131–141.

24 Stokłosa T, Gołąb J: Prospects for p53-based cancer therapy. Acta Biochim Pol 2005;2: 321–328.

25 Haupt S, Haupt Y: Improving cancer therapy through p53 management. Cell Cycle 2004;3: 912–916.

26 Zeimet AG, Marth C: Why did p53 gene therapy fail in ovarian cancer? Lancet Oncol 2003;4:415–422.

27 Guo J, Xin H: Chinese gene therapy. Splicing out the West? Science 2006;314:1232–1235.

28 Nemunaitis J, Clayman G, Agarwala SS, Hrushesky W, Wells JR, Moore C, Hamm J, Yoo G, Baselga J, Murphy BA, Menander KA, Licato LL, Chada S, Gibbons RD, Olivier M, Hainaut P, Roth JA, Sobol RE, Goodwin WJ: Biomarkers predict p53 gene therapy efficacy in recurrent squamous cell carcinoma of the head and neck. Clin Cancer Res 2009;15:7719–7725.

29 Roth JA, Nguyen D, Lawrence DD, Kemp BL, Carrasco CH, Ferson DZ, Hong WK, Komaki R, Lee JJ, Nesbitt JC, Pisters KM, Putnam JB, Schea R, Shin DM, Walsh GL, Dolormente MM, Han CI, Martin FD, Yen N, Xu K, Stephens LC, McDonnell TJ, Mukhopadhyay T, Cai D: Retrovirus-mediated wild-type p53 gene transfer to tumors of patients with lung cancer. Nat Med 1996;2:985–991.

30 Zeimet AG, Marth C: Why did p53 gene therapy fail in ovarian cancer? Lancet Oncol 2003;4:415–422.

31 Senzer N, Nemunaitis J, Nemunaitis M, Lamont J, Gore M, Gabra H, Eeles R, Sodha N, Lynch FJ, Zumstein LA, Menander KB, Sobol RE, Chada S: p53 therapy in a patient with Li-Fraumeni syndrome. Mol Cancer Ther 2007;6:1478–1482.

32 Pai SI, Lin Y-Y, Macaes B, Meneshian A, Hung CF, Wu TC: Prospects of RNA interference therapy for cancer. Gene Ther 2006;13:464–477.

33 Maen A, Stephen S, Cheryl B, Ala A: RNAi and cancer: implications and applications. J RNAi Gene Silencing 2006;2:136–145.

34 Wall NR, Shi Y: Small RNA: can RNA interference be exploited for therapy? Lancet 2003;362:1401–1403.

35 Mukhopadhyay T, Tainsky M, Cavender AC, Roth JA: Specific inhibition of k-ras expression and tumorigenicity of lung cancer cells by antisense RNA. Cancer Res 1991;51:1744–1748.

36 Shi XH, Liang ZY, Ren XY, Liu TH: Combined silencing of K-ras and Akt2 oncogenes achieves synergistic effects in inhibiting pancreatic cancer cell growth in vitro and in vivo combined silencing of K-ras and Akt2 oncogenes. Cancer Gene Ther 2009;16:227–236.

37 Réjiba S, Wack S, Aprahamian M, Hajri A: K-ras oncogene silencing strategy reduces tumor growth and enhances gemcitabine chemotherapy efficacy for pancreatic cancer treatment. Cancer Sci. 2007;98:1128–1136.

38 Scherer LJ, Rossi JJ: Approaches for the sequence-specific knockdown of mRNA. Nat Biotechnol 2003;21:1457–1465.

39 Kaufmann SH, Hengartner MO: Programmed cell death: alive and well in the new millennium. Trends Cell Biol 2001;11:526–534.

40 Dai Y, Lawrence TS, Xu L: Overcoming cancer therapy resistance by targeting inhibitors of apoptosis proteins and nuclear factor-κB. Am J Transl Res 2009;1:1–15.

41 Wack S, Rejiba S, Céline Parmentier C, Hajri A: Telomerase transcriptional targeting of inducible Bax/TRAIL gene therapy improves gemcitabine treatment of pancreatic cancer. Mol Ther 2008;16:252–260.

42 Ferreira CG, Epping M, Kruyt FAE, Giaccone G: Apoptosis: target of cancer therapy. Clin Cancer Res 2002;8:2024–2034.

43 Ivy PS, Schoenfeldt M: Clinical trials referral resource. Current clinical trials of 17-AAG and 17-DMAG. Oncology (Huntingt) 2004;18:610, 615, 619–620.

44 Mei S, Ho AD, Mahlknecht U: Role of histone deacetylase inhibitors in the treatment of cancer (review). Int J Oncol 2004;25:1509–1519.

45 Rinehart J, Adjei AA, Lorusso PM, Waterhouse D, Hecht JR, Natale RB, Hamid O, Varterasian M, Asbury P, Kaldjian EP, Gulyas S, Mitchell DY, Herrera R, Sebolt-Leopold JS, Meyer MB: Multicenter phase II study of the oral MEK inhibitor, CI-1040, in patients with advanced nonsmall cell lung, breast, colon, and pancreatic cancer. J Clin Oncol 2004;22:4456–4462.

46 Portsmouth D, Hlavaty J, Renner M: Suicide genes for cancer therapy. Mol Aspects Med 2007;28:4–41.

47 Li S, Tokuyama T, Yamamoto J, Koide M, Yokota N, Namba H: Bystander effect-mediated gene therapy of gliomas using genetically engineered neural stem cells. Cancer Gene Ther 2005;12:600–607.

48 Dachs GU, Hunt MA, Syddall S, Singleton DC, Patterson AV: Bystander or no bystander for gene directed enzyme prodrug therapy. Molecules 2009;14:4517–4545.

49 Hajri A, Wack S, Lehn P, Vigneron JP, Lehn JM, Marescaux J, Aprahamian M: Combined suicide gene therapy for pancreatic peritoneal carcinomatosis using BGTC liposomes. Cancer Gene Ther 2004;11:16–27.

50 Barton KN, Stricker H, Brown SL, Elshaikh M, Aref I, Lu M, Pegg J, Zhang Y, Karvelis KC, Siddiqui F, Kim JH, Freytag SO, Movsas B: Phase I trial of replication-competent adenovirus-mediated suicide gene therapy combined with IMRT for prostate cancer. Mol Ther 2008;16:1761–1769.

51 Xu F, Li S, Li XL, Guo Y, Zou BY, Xu R, Liao H, Zhao HY, Zhang Y, Guan ZZ, Zhang L: Phase I and biodistribution study of recombinant adenovirus vector-mediated herpes simplex virus thymidine kinase gene and ganciclovir administration in patients with head and neck cancer and other malignant tumors. Cancer Gene Ther 2009;16:723–30.

52 Deharvengt S, Wack S, Uhring M, Aprahamian M, Hajri A: Suicide gene/prodrug therapy for pancreatic adenocarcinoma by *E. coli* purine nucleoside phosphorylase and 6-methylpurine 2'-deoxyriboside. Pancreas 2004;28:E54–E64.

53 Tai CK, Wang W, Lai YH, Logg CR, Parker WB, Li YF, Hong JS, Sorscher EJ, Chen TC, Kasahara N: Enhanced efficiency of prodrug activation therapy by tumor-selective replicating retrovirus vectors armed with the *Escherichia coli* purine nucleoside phosphorylase gene. Cancer Gene Ther 2010;17:614–23.

54 Tedcastle A, Cawood R, Di Y, Fisher KD, Seymour LW: Virotherapy – cancer targeted pharmacology. Drug Discov Today 2011;17:215–220.

55 Meerani S, Yao Y: Oncolytic viruses in cancer therapy. Eur J Sci Res 2010;40:156–171.

56 Davis JJ, Fang B: Oncolytic virotherapy for cancer treatment: challenges and solutions. Gene Med 2005;7:1380–1389.

57 Wong HH, Lemoine NR, Wang Y: Oncolytic viruses for cancer therapy: overcoming the obstacles. Viruses 2010;2:78–106.

58 Cody JJ, Douglas JT: Armed replicating adenoviruses for cancer virotherapy. Cancer Gene Ther 2009;16:473–488.

59 Sova P, Ren XW, Ni S, Bernt KM, Mi J, Kiviat N, Lieber A: A tumor-targeted and conditionally replicating oncolytic adenovirus vector expressing TRAIL for treatment of liver metastases. Mol Ther 2004;9:496–509.

60 Fukazawa T, Matsuoka J, Yamatsuji T, Maeda Y, Durbin ML, Naomoto Y: Adenovirus-mediated cancer gene therapy and virotherapy. Int J Mol Med 2010;25:3–10.

Dr. Amor Hajri
Institute of Biology III, University of Freiburg
Schänzlestrasse 1
DE–79104 Freiburg (Germany)
Tel. +49 0 761 203 2762, amor.hajri@biologie.uni-freiburg.de

Piguet P, Poindron P (eds): Genetically Modified Organisms and Genetic Engineering in Research and Therapy. BioValley Monogr. Basel, Karger, 2012, vol 3, pp 103–119

Manipulating the Mitochondrial Genetic System

André Dietrich · Daria Mileshina · Adnan Khan Niazi · Anne Cosset · Romain Val · Noha Ibrahim · Frédérique Weber-Lotfi

Institut de Biologie Moléculaire des Plantes, CNRS and Université de Strasbourg, Strasbourg, France

Abstract

Cytoplasmic organelles in eukaryotes retain their own genome. Mutations in the human mitochondrial DNA are the most common cause of hereditary neuromuscular diseases. These degenerative disorders are still incurable and would need gene therapy. In plants, mitochondrial genetics provides traits of great agronomical relevance. Conventional methodologies failed to enable genetic transformation of mitochondria in mammalian and plant cells, but a variety of alternative complementation strategies and organelle transfection approaches are currently developed.

Genetic information in eukaryotic cells is distributed between the nucleus and cytoplasmic organelles, i.e. mitochondria in non-plant organisms, mitochondria and chloroplasts in plants. Mitochondria ensure fundamental functions in energy production, redox processes, metabolic pathways and cell death. Their genetic system provides a number of polypeptides which are essential for cell survival, as they are components of the respiratory chain or contribute to its biogenesis. All data imply that maintenance, integrity and efficient expression of the mitochondrial genome are fundamental for eukaryotic organisms. In human, mutations in the organelle genome cause severe neurodegenerative diseases that are currently incurable and await gene therapy strategies [1]. Mitochondrial genetics in plants influences a number of agronomically relevant traits [2]. 'Omics' approaches and functional analyses generate increasing new information in the mitochondrial field. However, filling the gaps in the understanding of the organelle genetic processes and complementing mitochondrial dysfunctions are restricted by the current inability to transform mitochondria in mammalian or plant cells. So far, only two unicellular organisms (yeast and the alga *Chlamydomonas reinhardtii*) are amenable to mitochondrial transformation, using biological ballistics methodologies [3, 4]. The challenge brought research groups to explore a number of alternative strategies.

Mitochondrial Genetic Systems

The mammalian mitochondrial DNA (mtDNA) is a small and compact circular molecule of about 16.5 kb. It carries 37 genes encoding 13 proteins of the oxidative phosphorylation complexes, 2 ribosomal RNAs (rRNAs) and 22 transfer RNAs (tRNAs). The information is distributed between the (+) strand (or heavy strand, 12 protein genes, 2 rRNA genes, 14 tRNA genes) and the (–) strand (or light strand, 1 protein gene, 8 tRNA genes). Mammalian mitochondrial genomes have no intergenic regions or introns and possess a single noncoding region carrying the transcription promoters (HSP1 and HSP2 on the heavy strand, LSP on the light strand) and one of the replication origins (O_H). Gene expression in these organelles involves a number of specific transcriptional and post-transcriptional steps [5] and might be partially regulated by nuclear transcription factors [6].

Higher plant mitochondria possess much larger genomes than mammalian organelles. These are commonly in the range of several hundred kb: 367 kb for the model plant *Arabidopsis thaliana* [7] but 740 kb for maize with the CMS-C cytoplasm [8]. Cucumber is so far the most extreme case, with a mitochondrial genome mapping as three circular chromosomes of 1,556, 84, and 45 kb [9]. Remarkably, these large genomes still contribute less than 60 identified genes [10], while containing long intergenic regions and introns. Further expansion comes from the acquisition of sequences from diverse sources, including nuclear and chloroplast genomes, viruses and bacteria. More than half of the sequences in plant mtDNAs are of unknown origin and function. Finally, in many plant species, mitochondria also possess one or several extrachromosomal plasmids of undetermined function and phylogenetic origin [11]. In addition, plant mitochondrial genomes exhibit a dynamic structure and the information is fractionated into different subgenomic forms resulting from homologous recombination between repeated regions, a process that is under nuclear control [12]. Along with a complex genome structure, plant mitochondrial gene expression mechanisms are particularly elaborate, including multiple promoters, *cis*- and *trans*-splicing of introns [13], RNA processing and surveillance [14], as well as extensive RNA editing [15].

    Dietrich · Mileshina · Niazi · Cosset · Val · Ibrahim · Weber-Lotfi

Human Mitochondrial Genetics and Neurodegenerative Diseases

Over 300 pathogenic mutations in the human mtDNA have been characterized (see http://www.mitomap.org). These fall into three types [1, 16]: (1) missense mutations in protein-coding genes, (2) point mutations in tRNA or rRNA genes that impair mitochondrial protein synthesis, and (3) duplications or deletions. mtDNA missense mutations can be associated with two common ophthalmologic manifestations, LHON (Leber's hereditary optic neuropathy) and NARP (neuropathy, ataxia and retinitis pigmentosa). The example diseases for tRNA point mutations are MERRF (myoclonic epilepsy and ragged-red fibers) and MELAS (mitochondrial encephalopathy, lactic acidosis and stroke-like episodes). mtDNA rearrangements can produce the Pearson syndrome (refractory sideroblastic anemia, pancytopenia, defective oxidative phosphorylation, exocrine pancreatic insufficiency, and variable hepatic, renal, and endocrine failure) or the CPEO syndrome (chronic progressive external ophthalmoplegia). CPEO may be part of KSS (Kearns-Sayre syndrome), a more severe and fatal multi-system disorder (muscle weakness, cerebellar damage, and heart failure). mtDNA deletions are a primary cause of mitochondrial disease and are likely to have a central role in the aging of postmitotic tissues. Large-scale mtDNA rearrangements occur as multiple short deletions or a single macrodeletion. Most are located in the major arc between two proposed origins of replication (O_H and O_L).

Mutations can affect all mtDNA copies in the cell (homoplasmic state) or only some copies (heteroplasmic state). Heteroplasmy is the most general situation and the onset of clinical symptoms is determined by the ratio of wild-type to mutant mtDNA, with a typical threshold effect [17]. This means that within a tissue there may be a mixture of respiratory-competent and deficient cells. Due to random sorting of mitochondria, heteroplasmy yields mitotic segregation, i.e. the proportion of mutated mtDNA copies may shift in daughter cells [18]. On the other hand, preferential amplification of the mutated mtDNA copies often occurs in post-mitotic tissues, leading to clonal expansion [19]. Finally, mtDNA genotypes may segregate between generations, suggesting the existence of an mtDNA bottleneck during development. The mechanism underlying such a process is still a matter of debate [1]. Hereditary mitochondrial diseases are transmitted only through the maternal line, since spermatozoa contain hardly any mitochondria. Paternal inheritance of an mtDNA disorder has been occasionally documented [20].

mtDNA deletions sufficient to generate mitochondrial dysfunction have been found in neurons of patients with Parkinson's disease [21] or multiple sclerosis [22]. Large scale mtDNA deletions also clonally expand in aging postmitotic tissues [23], whereas point mutations preferentially accumulate in aged mitotic tissues [24]. Whether mtDNA mutations can cause aging nevertheless remains to be clarified [24]. Finally, mtDNA mutations can promote tumor cell proliferation, support tumor adaptation to new environments or regulate metastasis [25, 26].

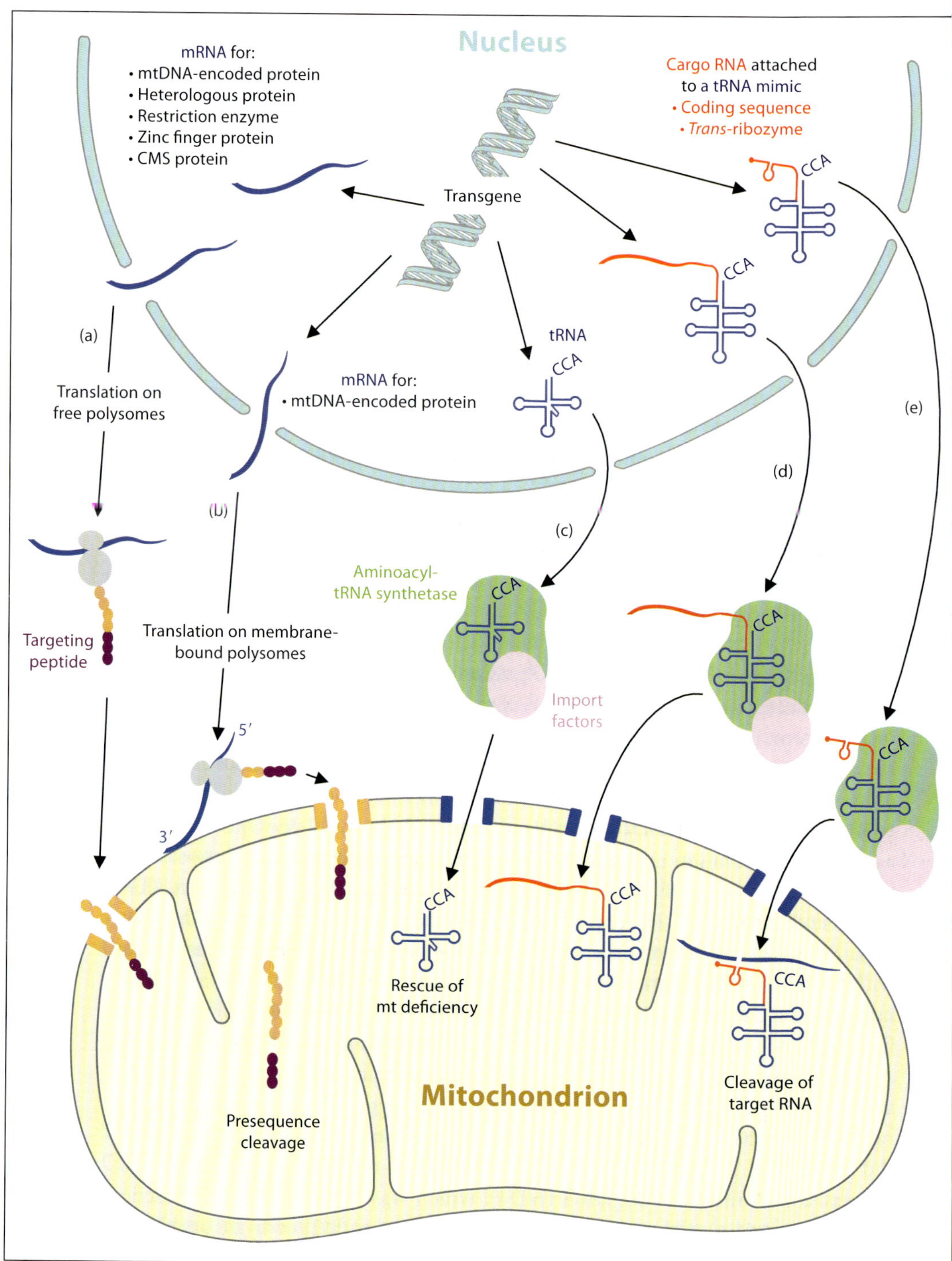

Nucleus
mRNA for:
• mtDNA-encoded protein
• Heterologous protein
• Restriction enzyme
• Zinc finger protein
• CMS protein
Cargo RNA attached to a tRNA mimic
• Coding sequence
• Trans-ribozyme
Transgene
CCA
CCA
tRNA
CCA
CCA
(a)
(b)
(c)
(d)
(e)
Translation on free polysomes
mRNA for:
• mtDNA-encoded protein
Translation on membrane-bound polysomes
Aminoacyl-tRNA synthetase
Import factors
Targeting peptide
CCA
CCA
CCA
5'
3'
CCA
CCA
CCA
Rescue of mt deficiency
CCA
Cleavage of target RNA
Presequence cleavage
Mitochondrion

established that the allotopically expressed protein was not translocated into mitochondria and that the rescued cells were revertants for the mutation [44].

A major difficulty in these approaches seems to be the high hydrophobicity of the proteins normally encoded by the mtDNA, which may hinder trafficking and organelle import. It has been established that a subclass of cytosolic mRNAs localizes to the mitochondrial surface in yeast and mammalian cells, which allows co-translational import of the corresponding proteins into the organelles [45, 46]. Exploiting this mechanism, Kaltimbacher et al. [47] showed that targeting the nuclear transgene-derived *atp6* transcript to the mitochondrial outer membrane facilitated co-translational translocation of the gene product, thus preventing the accumulation of cytosolic aggregates. A similar approach with nuclearly expressed *atp6*, *nd1* (encoding subunit 1 of complex I) and *nd4* transcripts led to long-lasting and complete rescue of mitochondrial dysfunction in human fibroblasts harboring the 8993T>G *atp6* mutation, the 3460G>A *nd1* mutation and the G11778G>A *nd4* mutation, respectively [48, 49]. In further experiments, the strategy was validated in a rat model of mitochondrial dysfunction [50]. In summary, allotopic expression of mitochondrial proteins in mammalian cells has generated some apparently positive results but in a number of reports analysis showing unequivocally that the protein expressed from the nuclear transgene is indeed imported into the organelles and assembled into the relevant mitochondrial complex is lacking [44].

Nuclear Expression and Mitochondrial Import of Heterologous Proteins in Mammals

Finally, other strategies have exploited genes that are normally absent in human but that can bypass nonfunctional steps in the respiratory chain. The single subunit NADH dehydrogenase encoded by the *S. cerevisiae* NDI1 gene transfers electrons from mitochondrial NADH to the quinone pool. When introduced into heterologous mitochondria, it can bypass complex I. The yeast Ndi1 protein expressed from a nuclear transgene in mammalian cell lines was imported into mitochondria and restored the NADH oxidase activity of complex I-deficient

Fig. 1. Schematic view of nuclear transgenesis-based strategies to target proteins and RNAs of interest into mitochondria. Protein genes normally carried by the mtDNA and genes encoding heterologous proteins, restriction endonucleases, specific zinc finger proteins or CMS proteins were expressed in the nucleus after addition of a sequence coding for a mitochondrial targeting peptide; the mRNAs were either translated on free cytosolic polysomes (a) or specifically targeted to the mitochondrial surface (b), leading to membrane-bound polysomes; translocation into the organelles was through the regular protein import pathway. Genes encoding tRNA variants (c) and gene constructs encoding cargo RNAs (coding sequence or *trans*-cleaving ribozyme) attached to a tRNA mimic (d and e) were expressed in the nucleus; the resulting transcripts associated with the specific aminoacyl-tRNA synthetase and import factors and were translocated into mitochondria through the natural tRNA import pathway.

cells [51, 52]. To test the therapeutic prospects of this approach, complementation of complex I deficiency by the yeast Ndi1 enzyme was validated in a mouse model of Parkinson's disease [53]. Further, plants and some animals possess an alternative oxidase that transfers electrons from the quinone pool to oxygen, bypassing the whole cytochrome pathway and generating water (e.g. [54]). Nuclear expression and mitochondrial targeting of *Ciona intestinalis* alternative oxidase was shown to complement cytochrome oxidase deficiency in human cells [55].

Nuclear Expression and Mitochondrial Import of Antigenomic Enzymes in Mammals

As mentioned above, pathogenic mtDNA mutations are most often heteroplasmic, i.e. wild-type and mutated DNA molecules co-exist. The appearance and severity of disease symptoms depend on the relative proportion of mutated over wild-type mtDNA copies. The threshold can be over 90% of mutated copies. 'Antigenomic' strategies have therefore been developed to promote elimination of the mutated DNA molecules and shift the heteroplasmy in favor of the wild-type genomes. These approaches consisted of nuclear expression and mitochondrial import of enzymes directed against the mutated mtDNA. Organelle import of the *Pst*I restriction endonuclease driven by a mitochondrial targeting peptide-triggered degradation of mtDNA copies harboring the corresponding sites in mammalian cells and enabled to modulate heteroplasmy [56]. Nuclear expression and mitochondrial uptake of further relevant endonucleases in mammalian cell lines or animal models subsequently provided specific cleavage of restriction sites created by pathogenic mutations in the mtDNA [57–60]. Yet, this strategy has limitations. Expression of restriction endonucleases may generate nuclear DNA damage and the strategy only applies when the mutation under scrutiny creates a suitable restriction site in the mtDNA. These limitations may be circumvented by combining the enzyme of interest with zinc finger peptides that will bind with high specificity to predetermined sequences in the mtDNA and discriminate between mutated and wild-type motifs. Mitochondrial targeting of a methylase combined with relevant zinc finger peptides in human cells carrying the 8993T>G mtDNA mutation resulted in selective methylation of cytosines adjacent to the mutation site [61]. Moreover, when a nuclease domain was fused to the specific zinc fingers, the mtDNA was cleaved at predicted sites adjacent to the mutation, leading to a selective degradation of the mutated mtDNA copies and a shift of the heteroplasmy towards the wild-type genomes [62].

Nuclear Expression and Mitochondrial Import of tRNAs in Human Cells

As many disorders of the mitochondrial genome are caused by mutations in mitochondrial tRNA-encoding genes, there has been special interest in rescuing these mutations by importing normal copies of the relevant tRNA from the cytosol. Mitochondrial import of nuclear-encoded tRNAs is a process occurring

      Dietrich · Mileshina · Niazi · Cosset · Val · Ibrahim · Weber-Lotfi in many organisms, mostly to compensate for the absence of some or even all tRNA genes in the mtDNA [63]. Although possessing an apparently complete set of tRNA genes, human mitochondria proved able to import yeast or human tRNA^{Lys} variants in vitro [64]. Mitochondrial import of tRNA^{Gln} has also been implied in human cells [65]. Indeed, Kolesnikova et al. [66] showed that yeast tRNA^{Lys} variants expressed from nuclear transgenes in human cell lines carrying the tRNA^{Lys} 8344A>G MERRF mutation were imported into the mitochondria and partially rescued the mitochondrial functional deficiency. Similarly, nuclear expression of recombinant tRNAs with a leucine identity in human cell lines allowed a significant rescue of the mitochondrial tRNA^{Leu} 3243A>G mutation, the major cause of the MELAS syndrome [67].

Transgenic Strategies to Manipulate Mitochondrial Genetics in Plants

Nuclear Expression of Mitochondrial CMS Genes in Plants
CMS is a trait of great relevance in hybrid production and crop breeding. Allotopic expression in plants has thus been used originally in attempts to produce directed male sterility by nuclear transformation with CMS sequences. As mentioned above, in a number of natural CMS systems sterility was shown to be associated with abnormal mitochondrial genes expressing aberrant proteins [32]. This was thought to provide an opportunity to induce male sterility by the production of nuclear transformants carrying transgenes that encode CMS proteins fused to a mitochondrial targeting peptide (fig. 1). However, the only reported example of successful production of male sterile plants by nuclear expression of a CMS gene concerns the *orf239* gene from bean [68]. Similar attempts with the proteins associated with maize CMS-T, petunia CMS or *Raphanus* and *Brassica* Ogura CMS did not give rise to sterile plants [69–72].

Nuclear Expression and Mitochondrial Import of Customized RNAs in Plants
In higher plants, mitochondria import one third to one half of their tRNAs from the cytosol [73]. Exploiting this uptake pathway allowed to develop a shuttle system for mitochondrial import of customized cargo RNAs expressed from nuclear transgenes in plant cells (fig. 1). It was established that a tRNA mimic, the tRNA-like structure derived from the 3′ end of the *Turnip yellow mosaic virus* genomic RNA [74], can drive a 5′ trailor cargo RNA into mitochondria in plant cell lines and whole plants [75]. This strategy opened the way for RNA-directed manipulation of the mitochondrial genetic system for both fundamental investigations and biotechnological applications. A chimeric transcript made of a *trans*-cleaving hammerhead ribozyme attached to the 5′ end of the tRNA-like shuttle was subsequently expressed in transformed plants and imported into the organelles, leading to the first directed knockdown of a mitochondrial RNA in a multicellular eukaryote [75]. Such an approach might in the future be

Table 1. Different strategies used to transfer DNA into mitochondria from various organisms

Mitochondrial transfection strategies	Organisms	References
DNA transfer into mitochondria in vitro		
Conjugation of mitochondria with bacteria carrying a DNA construct with an *oriT* element	mammal	76
Import of DNA covalently linked to a mitochondrial targeting peptide	yeast, mammal	77, 78, 79
Translocation upon electroporation	protist, mammal, plant	80, 81, 82, 83, 84
Incubation of linear DNA with mitochondria in relevant conditions	plant, mammal, yeast	85, 86, 87
DNA transfer into mitochondria in whole cells		
Transfer of intact mitochondria into recipient cells		
Use of cybrid or microinjection of mitochondria	mammal	88
Co-incubation of mitochondria with ρ^0 cells	mammal	89
Biological ballistics	green alga, yeast	3, 4
Exposure of cells to DNA-loaded DQAsomes or DNA-loaded mitochondriotropic liposomes	mammal	90, 91, 92
Protofection	mammal	93, 94

used to identify new functions in the large unassigned regions of the plant mitochondrial genomes or to trigger RNA-mediated male sterility.

Development of Mitochondrial Transfection Strategies

In parallel with the above strategies based on nuclear expression of transgenes and mitochondrial targeting of the resulting products, a variety of approaches have been explored to develop DNA transfer into mitochondria in vitro or in intact cells (table 1). Although mitochondrial transformation per se still has not been accessed, ideas and functional investigations are progressing.

Conjugation of isolated mammalian mitochondria with recombinant bacteria carrying a DNA construct with an origin of transfer *(oriT)* allowed to translocate the construct into the organelles [76]. Other studies have attempted to exploit the mitochondrial protein import pathway to translocate DNA. Single-stranded or double-stranded DNA covalently linked to a mitochondrial precursor protein or a mitochondrial targeting peptide was reportedly taken up into isolated yeast [77] or mammalian [78, 79] mitochondria. DNA has also been introduced into isolated trypanosomatid [80], mammalian [81, 82] and plant

[83, 84] mitochondria by electroporation. Remarkably, further studies established that isolated plant, mammalian or yeast mitochondria are actually competent to import linear DNA, through a presumed physiological mechanism [85–87]. In most of these approaches, the exogenous DNA introduced into the mitochondria can be expressed in the organelles [76, 80, 82–86].

Cell biology approaches based on the transfer of whole mitochondria into recipient cells have been explored in mammals, including cybrid formation and microinjection [88]. Furthermore, it has been reported that mammalian cells can ultimately take up isolated mitochondria [89]. However, application of such approaches to gene therapy would be extremely complex and further studies aimed to develop 'vehicles' able to deliver DNA into the cells and to the mitochondria.

As mentioned in the introduction, mitochondrial transfection through biological ballistics, i.e. bombardment of the cells with DNA-coated tungsten or gold particles, has been successful only in yeast and *C. reinhardtii* [3, 4]. This approach has failed so far with animal and higher plant cells, possibly due to the small organelle size or the absence of appropriate functional selection. Dequalinium-based cationic vesicles (so-called 'DQAsomes') and mitochondriotropic liposomes proved capable to deliver DNA into mammalian cells and to drive their cargo towards mitochondria [90, 91]. Further studies reported DQAsome-mediated mitochondrial delivery of a DNA construct containing the green fluorescent protein gene in mammalian cell lines [92]. Although the delivery efficiency was low, expression of the GFP reporter protein was detected. Finally, a strategy called 'protofection' has been documented [93, 94]. To carry the DNA into the cells and to the mitochondria, it relies on a vehicle that combines a protein transduction domain and a mitochondrial targeting peptide with the TFAM mitochondrial transcription/mtDNA-binding factor. Rescue of mitochondrial functions in a Parkinson's disease cybrid model was reported [93], but surprisingly the vehicle without a DNA load triggered a similar response in this context [94].

Further Prospects

Mitochondrial transfection and rescue of mtDNA pathogenic mutations are still the subject of cutting edge research. Further potential applications of the zinc finger technology for targeting and modifying the mitochondrial genome have been proposed [95]. On the other hand, more strategies for organelle targeting of nuclearly-expressed RNAs are likely to become available in the next future. Mammalian mitochondria also import the 5S rRNA [64]. The structural elements allowing the uptake have been characterized, opening the possibility to target into the organelles mutated 5S rRNA versions in which domains dispensable for import have been substituted with sequences of interest [96]. Furthermore, a set of mitochondrially importable small tRNA-derived aptamers has been selected in vitro and these might become suitable as well to target customized RNA oligomers into deficient human mitochondria [97]. Also, stem-loop structures that function as mitochondrial RNA targeting signals were identified in the mammalian RNase P RNA and MRP RNA through in vitro import assays [98]. They may in turn be used to promote organelle translocation of RNAs of interest. In other assays, efficient internalization of various RNAs into isolated plant, yeast or mammalian mitochondria was obtained in the presence of a shuttle protein carrying an organelle targeting signal, a strategy potentially suitable for the translocation of mRNAs [99]. Finally, further studies reported cell uptake and mitochondrial transport of RNA mediated by a multi-subunit complex isolated from the protozoal parasite *Leishmania tropica*. Organelle targeting of a polycistronic mRNA appeared to rescue mitochondrial functions in a cybrid cell line derived from a patient with Kearns Sayre syndrome (KSS) and carrying an mtDNA deletion [100]. As to mitochondrial targeting of DNA, a further type of liposome-based nanocarrier for macromolecule delivery, the MITO-Porters, is currently developed [101]. Such particles can fuse with mitochondria and release their cargo into the organelles. Whether MITO-Porters can accommodate and deliver DNA remains to be established but it was reported that they are able to mediate the transfer of the GFP reporter protein or the DNase I nuclease into the mitochondria of human cells [101, 102].

Acknowledgments

Our projects are funded by the French Centre National de la Recherche Scientifique (CNRS, UPR2357), the Université de Strasbourg (UdS) and the Agence Nationale de la Recherche (ANR-06-MRAR-037–02, ANR-09-BLAN-0240–01).

References

1   Tuppen HA, Blakely EL, Turnbull DM, Taylor RW: Mitochondrial DNA mutations and human disease. Biochim Biophys Acta 2010;1797:113–128.

2   Frei U, Peiretti EG, Wenzel G: Significance of cytoplasmic DNA in plant breeding; in Janick J (ed.): Plant Breeding Reviews. Hoboken, John Wiley & Sons, 2004, pp 175–210.

3   Bonnefoy N, Remacle C, Fox TD: Genetic transformation of *Saccharomyces cerevisiae* and *Chlamydomonas reinhardtii* mitochondria. Methods Cell Biol 2007;80:525–548.

4   Zhou J, Liu L, Chen J: Mitochondrial DNA heteroplasmy in *Candida glabrata* after mitochondrial transformation. Eukaryot Cell 2010;9:806–814.

5   Falkenberg M, Larsson NG, Gustafsson CM: DNA replication and transcription in mammalian mitochondria. Annu Rev Biochem 2007;76:679–699.

6   Leigh-Brown S, Enriquez JA, Odom DT: Nuclear transcription factors in mammalian mitochondria. Genome Biol 2010;11:215.

7 Unseld M, Marienfeld JR, Brandt P, Brennicke A: The mitochondrial genome of *Arabidopsis thaliana* contains 57 genes in 366,924 nucleotides. Nat Genet 1997;15:57–61.

8 Allen JO, Fauron CM, Minx P, Roark L, Oddiraju S, Lin GN, Meyer L, Sun H, Kim K, Wang C, Du F, Xu D, Gibson M, Cifrese J, Clifton SW, Newton KJ: Comparisons among two fertile and three male-sterile mitochondrial genomes of maize. Genetics 2007;177:1173–1192.

9 Alverson AJ, Rice DW, Dickinson S, Barry K, Palmer JD: Origins and recombination of the bacterial-sized multichromosomal mitochondrial genome of cucumber. Plant Cell 2011;23:2499–2513.

10 Kubo T, Newton KJ: Angiosperm mitochondrial genomes and mutations. Mitochondrion 2008;8:5–14.

11 Handa H: Linear plasmids in plant mitochondria: peaceful coexistences or malicious invasions? Mitochondrion 2008;8:15–25.

12 Maréchal A, Brisson N: Recombination and the maintenance of plant organelle genome stability. New Phytol 2010;186:299–317.

13 Bonen L: *Cis-* and *trans-*splicing of group II introns in plant mitochondria. Mitochondrion 2008;8:26–34.

14 Holec S, Lange H, Canaday J, Gagliardi D: Coping with cryptic and defective transcripts in plant mitochondria. Biochim Biophys Acta 2008;1779:566–573.

15 Chateigner-Boutin AL, Small I: Organellar RNA editing. WIREs RNA 2011;2:493–506.

16 Schon EA, DiMauro S: Mitochondrial mutations: genotype to phenotype. Novartis Found Symp 2007;287:214–233.

17 Wong LJ: Diagnostic challenges of mitochondrial DNA disorders. Mitochondrion 2007;7:45–52.

18 Rahman S, Poulton J, Marchington D, Suomalainen A: Decrease of 3243 A–>G mtDNA mutation from blood in MELAS syndrome: a longitudinal study. Am J Hum Genet 2001;68:238–240.

19 Elson JL, Samuels DC, Turnbull DM, Chinnery PF: Random intracellular drift explains the clonal expansion of mitochondrial DNA mutations with age. Am J Hum Genet 2001;68:802–806.

20 Schwartz M, Vissing J: Paternal inheritance of mitochondrial DNA. N Engl J Med 2002;347:576–580.

21 Bender A, Krishnan KJ, Morris CM, Taylor GA, Reeve AK, Perry RH, Jaros E, Hersheson JS, Betts J, Klopstock T, Taylor RW, Turnbull DM: High levels of mitochondrial DNA deletions in substantia nigra neurons in aging and Parkinson disease. Nat Genet 2006;38:515–517.

22 Campbell GR, Ziabreva I, Reeve AK, Krishnan KJ, Reynolds R, Howell O, Lassmann H, Turnbull DM, Mahad DJ: Mitochondrial DNA deletions and neurodegeneration in multiple sclerosis. Ann Neurol 2011;69:481–492.

23 Kraytsberg Y, Kudryavtseva E, McKee AC, Geula C, Kowall NW, Khrapko K: Mitochondrial DNA deletions are abundant and cause functional impairment in aged human substantia nigra neurons. Nat Genet 2006;38:518–520.

24 Greaves LC, Reeve AK, Taylor RW, Turnbull DM: Mitochondrial DNA and disease. J Pathol 2011:doi: 10.1002/path.3028.

25 Brandon M, Baldi P, Wallace DC: Mitochondrial mutations in cancer. Oncogene 2006;25:4647–4662.

26 Imanishi H, Hattori K, Wada R, Ishikawa K, Fukuda S, Takenaga K, Nakada K, Hayashi J: Mitochondrial DNA mutations regulate metastasis of human breast cancer cells. PLoS One 2011;6:e23401.

27 Ducos E, Touzet P, Boutry M: The male sterile G cytoplasm of wild beet displays modified mitochondrial respiratory complexes. Plant J 2001;26:171–180.

28 Karpova OV, Kuzmin EV, Elthon TE, Newton KJ: Differential expression of alternative oxidase genes in maize mitochondrial mutants. Plant Cell 2002;14:3271–3284.

29 Lilly JW, Bartoszewski G, Malepszy S, Havey MJ: A major deletion in the cucumber mitochondrial genome sorts with the MSC phenotype. Curr Genet 2001;40:144–151.

30 Chase CD: Cytoplasmic male sterility: a window to the world of plant mitochondrial-nuclear interactions. Trends Genet 2007;23:81–90.

31 Carlsson J, Leino M, Sohlberg J, Sundstrom JF, Glimelius K: Mitochondrial regulation of flower development. Mitochondrion 2008;8:74–86.